# Inorganic Chemistry
## Introduction to Coordination Chemistry

Edward Lisic, Ph.D.

All rights reserved. No part of this book shall be reproduced or transmitted in any form or by any means, electronic, mechanical, magnetic, photographic including photocopying, recording or by any information storage and retrieval system, without prior written permission of the publisher. No patent liability is assumed with respect to the use of the information contained herein. Although every precaution has been taken in the preparation of this book, the publisher and author assume no responsibility for errors or omissions. Neither is any liability assumed for damages resulting from the use of the information contained herein.

Copyright © 2012 by Edward Lisic, Ph.D.

ISBN 978-0-7414-7857-3 Paperback
ISBN 978-0-7414-7858-0 Hardcover
ISBN 978-0-7414-7859-7 eBook

Printed in the United States of America

Published July 2012

INFINITY PUBLISHING
1094 New DeHaven Street, Suite 100
West Conshohocken, PA 19428-2713
Toll-free (877) BUY BOOK
Local Phone (610) 941-9999
Fax (610) 941-9959
Info@buybooksontheweb.com
www.buybooksontheweb.com

*This book is dedicated to
anyone and everyone who has ever
synthesized a new ligand--- and then
watched it react with a transition metal.*

# Preface

This book was written for a broad introductory sophomore inorganic class, CHEM 2010, which is one of our "Foundation" courses. The only other inorganic chemistry texts were huge tomes and prohibitively expensive for a small enrollment course. I wanted a broad text on introductory transition metal chemistry, yet I wanted to impart to my students a love of the structures of these unique complexes and the flavor of their reactions with strange new ligands.

This book expands yet concisely clarifies material taught in Freshman Chemistry, as well as adding the descriptive chemistry of coordination chemistry and re-introducing the student to molecular orbital theory. The chapter 8 bonding in solids material can be taught along with the material in chapter 1, yet I felt it was better to cover molecular orbital theory before covering solids.

This text, in my opinion, is perfect for a sophomore level Transition Metal class, for a freshman Honors class, a freshman inorganic chemistry class, or as an inexpensive and concise supplement for advanced courses.

*July $24^{th}$, 2012     Edward C. Lisic*

# Table of Contents

✓ **Chapter 1.**  p 1
**Introduction to Inorganic Chemistry**

**Chapter 2.**  p 45
**Lewis Electron Dot Structures and Valence Shell Electron Pair Repulsion Theory**

✓ **Chapter 3.**  p 85
**Ligands and Nomenclature**

**Chapter 4.**  p 118
**Stereochemistry**

✓ **Chapter 5.**  p 134
**Crystal Field Theory**

✓ **Chapter 6.**  p 170
**Molecular Orbital Theory**

**Chapter 7.**  p 193
**Organometallics**

**Chapter 8.**  p 216
**Bonding in Solids**

**History of Transition Metal Inorganic Chemistry**  p 233

**Answers to Problems and Exercises**  p 240

# Table of Contents

Chapter 1.  
Introduction to Inorganic Chemistry — p 1

Chapter 2.  
Lewis Electron Dot Structures and Valence Shell Electron Pair Repulsion Theory — p 45

Chapter 3.  
Ligands and Nomenclature — p 85

Chapter 4.  
Stereochemistry — p 115

Chapter 5.  
Crystal Field Theory — p 134

Chapter 6.  
Molecular Orbital Theory — p 170

Chapter 7.  
Organometallics — p 193

Chapter 8.  
Bonding in Solids — p 216

History of Transition Metal Inorganic Chemistry — p 233

Answers to Problems and Exercises — p 240

# Chapter 1. Introduction to Inorganic Chemistry

Chemistry can be loosely classified into two major categories; (1) organic chemistry and (2) inorganic chemistry. Organic chemistry is an area of chemistry which involves the study of carbon-based compounds, hydrocarbons, and their derivatives. On the other hand, inorganic chemistry is an area of chemistry which involves the study of compounds of all the other elements. One view of inorganic compounds is that they have no biological origin, and another is that they can be classified as what they are not---they are not organic compounds.

Descriptive inorganic chemistry focuses on the classification of inorganic compounds. This is a complicated and difficult task due to the widely varying chemical properties of so many elements. One of the complicating factors involves chemical bonding of the elements. Usually, students think of only ionic and covalent bonding as the primary types of chemical bonds, but inorganic chemists understand that metallic bonding is very important in understanding the physical properties of compounds.

Chemical bonding of the elements helps us to understand the different types of intramolecular bonding between elements in compounds, but there are also additional intermolecular forces, such as Van der Waals forces that must also be recognized as conferring significant properties to solids. Classifications can be determined by comparing and observing properties of the compounds, or by determining such things as the heaviest element in the compound, the central element in the compound, or the overall structural type.

Centuries ago, alchemists listed the three alchemical principles of the world and designated three substances to exemplify these three principles. Even though our science is wholly more advanced than the alchemists of medieval times, it is somewhat instructive to understand the tainted history of chemistry.

The three alchemical principles were exemplified by mercury, sulfur, and salt, shown in the scheme below.

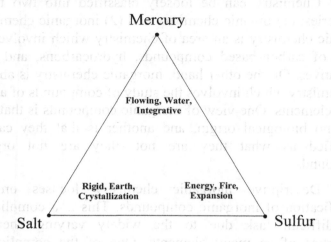

These three alchemical principles were very loose and general concepts that gave the notion that materials weren't inanimate objects, but that they had properties beyond what we modern scientists would confer on them. But, before we discard the old alchemical triangle as mystical nonsense, let's take a moment to consider the substances themselves.

Salt is an ionic substance, typically NaCl, whereas sulfur is actually a covalent molecular substance, usually found in nature as molecular $S_8$. When we realize that the third member of the triangle is a metallic substance, then it seems to me that we have found something fundamentally important in the old alchemist's triangle. These could actually represent, in modern terms, ionic compounds, covalent compounds, and metals.

These three bonding types can be represented with their own triangle. In the modern literature we have several examples of this type of bonding classification representation as anticipated in 1928 by Grimm, and then first represented in 1935 by Fernelius and Robey.

Fernelius & Robey's bond-type triangle was published and shown pictorially in Fernelius, W.C. and Robey, R.F. The Nature of the Metallic State. J. Chem. Educ. 1935, 12, 53–68, and it shows something reminiscent of the alchemical triangle with appropriate modern interpretations of chemical bonding.

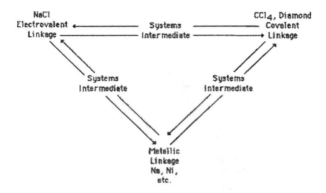

(scan pictures courtesy of Mark Leach, the Chemogenesis Web Project)

In 1941, van Arkel recognized the three types of extreme bonding and suggested another triangle to depict the bonding types, with intermediate compounds in between the extremes of the three bonding types.

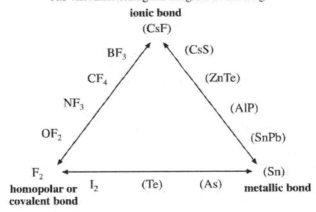

van Arkel, A.E., *Molecules and Crystals*, Butterworths, London, 1949, p205
first published in Dutch in 1941
(after Jensen *J.Chem.Educ.* (1995), **72**, 395)

Mark Leach and others have used the different versions of the van Arkel triangle using element combinations of the second period.

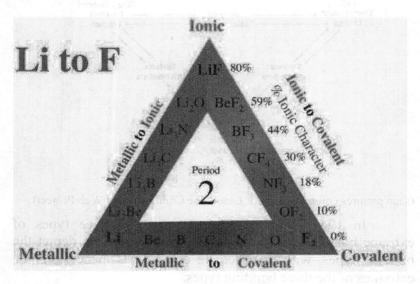

Courtesy of Mark Leach, Chemogenesis Webbook at:
http://www.meta-synthesis.com/webbook/37_ak/triangles.html

The Leach triangle shows students of chemistry an important concept: extremes in bonding are rare in chemistry, almost all compounds show some percentage or amount of different types of chemical bonds, and the lines between each type of bonding are blurred and overlap.

## Types of Crystalline Inorganic Solids

In one classification scheme, there are really four types of crystalline homogeneous inorganic compounds:

(1) Ionic compounds

(2) Covalent Main-Group compounds

(3) Metals

(4) Network solids.

This text will address the first three types of inorganic compounds in some depth, but solid-state network solids---the

# Inorganic Chemistry: Introduction to Coordination Chemistry

area of material science---is an advanced topic better left to the student to explore at their own pace or in an advanced class. The focus of this book, however, is to be an introduction to inorganic chemistry, and in particular, coordination chemistry.

To adequately describe and evaluate coordination chemistry it is absolutely necessary to discuss the chemistry of metals, ionic compounds and main group compounds first. Coordination compounds are metal-centered, and can be molecules and/or polyatomic ionic solids. Characteristically they exhibit coordination of electron donating ligands around the metal center.

This first chapter will deal with metals and ionic compounds, and the next two chapters will describe main group compounds and their special relationship to coordination chemistry.

## Metals

Typically the study of metals and their alloys (metallurgy) is not considered to be a topic of inorganic chemistry, since these are usually pure elements or homogeneous mixtures of elements (such as in sterling silver or brass). However, this is sometimes disputed, and the production of pure metals from ores is an active area of industrial chemistry, and the surface chemistry of metals is an active area of research in catalysis.

What is a metal? What are their characteristics? Typically in general chemistry texts a metal has (1) luster (2) malleable (3) ductile, and (4) good conductors of heat and electricity. But some non-metals, such as crystalline $I_2$ has luster, and so does fool's gold (iron pyrite $FeS_2$) so that isn't a good guide, and some metals are fairly brittle and aren't all that malleable or ductile.

The best criterion for a metal is that it is electrically conductive in three dimensions at standard ambient temperature and pressure, which are 25° C and 100 kPa pressure. Tin at temperatures below 18° C exists in a stable allotrope that is non-electrically conducting and brittle (ask Napoleon's soldiers). A more specific criterion is the temperature dependence of electrical conductivity; a metal has decreasing conductivity with increasing temperature.

Metals are found to the left of the seven semi-metals (or they are sometimes called metalloids) in the periodic table. These seven semi-metals are: Boron (B), Silicon (Si), Germanium (Ge), Arsenic (As), Antimony (Sb), Tellurium (Te), and Polonium (Po), although many do not consider Po to be a semi-metal since it is a radioactive element.

Metal atoms typically have large atomic radii. The effective nuclear charge that is felt by valence electrons in metal atoms is small compared to non-metals in the same period, thus the atomic radius is larger for a metal in the same period with a non-metal.

For example, as can be seen in on the next page, an atom of lithium, the lightest and least dense of the metals, has an atomic radius of 152 picometers, whereas the atomic radius of fluorine is only 64 picometers. That is astounding! The fact that fluorine is almost one third the size of lithium is counter-intuitive. Intuitively students will expect that an atom like fluorine, which has 9 protons, nine neutrons, and nine electrons, would be larger than an atom of lithium, which has only three protons, etc. To the un-initiated, lithium should be one third the size of fluorine...not the other way around!

This size difference is due to the effective nuclear charge that the valence electrons experience around the atom. Valence electrons do not shield each other very well from the attraction of the positive nuclear charge from the protons in the nucleus, and therefore as we go across a period, the size goes down! Valence electrons of nonmetals experience a high effective nuclear charge, and thus are pulled closer to the nucleus making their atoms smaller.

Semi-metals are smaller than metals, and non-metal atoms are the smallest.

# Inorganic Chemistry: Introduction to Coordination Chemistry

**Figure 1.1 Atomic radii of the Main-Group elements.**

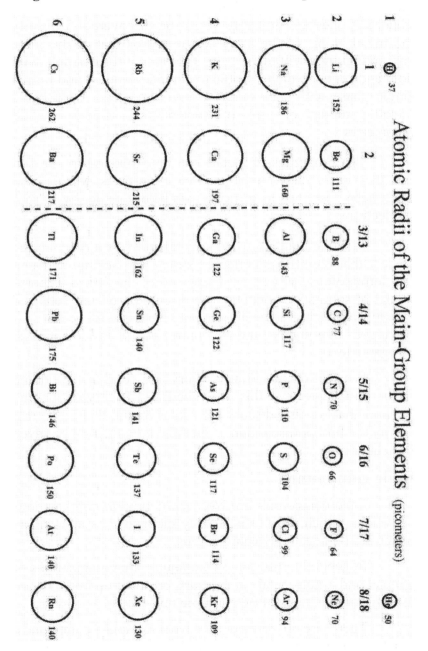

Metal alloys are important in modern technology, and thousands of alloy combinations are know, and many important ones are probably yet to be discovered. Typically the metal atoms of the alloy are of approximately the same size so that they will mix freely. This is true for example with gold and copper which are able to completely alloy and form a single phase with each other in any percentage. Gold has a metallic radius of 144 picometers and copper is 128 picometers.

The properties of metals can be explained by the fact that they have freely movable valence electrons. Metals have often been described as positive ions held together by a "sea" of electrons.

Metallic bonding in bulk metal solids is a complicated subject that requires attention to structure, packing, and unit cell terminology, as well as discussion of the "Band Theory" model of solid-state bonding. The discussion of unit cells and packing is touched upon in general chemistry texts, but it is discussed thoroughly and in more detail in senior-level inorganic texts.

We will discuss Band Theory and solid-state bonding further in Chapter 8 of this text. The Band Theory of bonding requires that we first discuss molecular orbital theory, so that is why the further discussion of metallic bonding will be saved for a later time.

## Ionic compounds

Ionic compounds are *electrically neutral* combinations of cations which are (+) positively charged ions and anions which are (-) negatively charged ions.

In general terms, ionic compounds are combinations of a metal and a non-metal. In contrast, covalent compounds are usually combinations of two or more non-metals.

There are in general two types of ionic compounds. Type I ionic compounds have metals that form only one type of cation. Type II ionic compounds have transition metals or heavy main group metals as cations, and the metal present can form two or more cations that have different charges. The nomenclature rules of Type I and Type II are slightly different, and we will deal with these types one at a time.

**Type I ionic compounds**

The most common example of a Type I ionic compound is table salt, which is sodium chloride (NaCl). It is the simplest example of a metal + non-metal combination.

Examples of some common simple ions are given in Table 1.1. In writing the correct formulas of such ionic compounds, the symbol of the metal cation is always given first and then followed by the symbol of the non-metal anion.

In naming ionic compounds of Type I, the metallic element is named first and keeps it elemental name, and the non-metal element is named last, and the element name for the anion keeps the root elemental name but given an -*ide* suffix. Thus NaCl is named sodium chlor*ide*, not sodium chlor*ine*. More examples are given in Table 1.2.

The reason that ionic compounds are usually metal plus non-metal combinations is due to the characteristic periodic trends of the metals and the non-metals.

Metallic elements tend to lose electrons and form cations, whereas non-metal elements tend to gain electrons and form anions. The driving force for this chemical change is so that the elements can achieve a noble gas-like electron configuration.

The electrical charge an ion exhibits is related to the Group to which that element belongs in the periodic table. This can be illustrated by looking at the metal and the non-metal sections of the main group elements of the periodic table and see the ions they form, as illustrated in Figure 1.2.

**Figure 1.2** Group Trends in Ion Formation

| 1 | 2 | | | 13 | 14 | 15 | 16 | 17 | 18 |
|---|---|---|---|---|---|---|---|---|---|
| $H^+$ | | | | | | | | | He |
| $Li^+$ | $Be^{2+}$ | | | B | C | $N^{3-}$ | $O^{2-}$ | $F^-$ | Ne |
| $Na^+$ | $Mg^{2+}$ | 3 | 12 | $Al^{3+}$ | Si | $P^{3-}$ | $S^{2-}$ | $Cl^-$ | Ar |
| $K^+$ | $Ca^{2+}$ | $Sc^{3+}$ | $Zn^{2+}$ | $Ga^{3+}$ | Ge | As | $Se^{2-}$ | $Br^-$ | Kr |
| $Rb^+$ | $Sr^{2+}$ | $Y^{3+}$ | $Cd^{2+}$ | In | Sn | Sb | Te | $I^-$ | Xe |
| $Cs^+$ | $Ba^{2+}$ | $La^{3+}$ | | | | | | | |

The following table lists some common ions that are formed from main-group elements in the periodic table.

**Table 1.1** Some Common Simple Cations and Anions.

| Cation | Name | Anion | Name |
|---|---|---|---|
| $H^+$ | Hydrogen | $H^-$ | hyd*ride* |
| $Li^+$ | Lithium | $F^-$ | fluo*ride* |
| $Na^+$ | Sodium | $Cl^-$ | chlo*ride* |
| $K^+$ | Potassium | $Br^-$ | brom*ide* |
| $Mg^{2+}$ | Magnesium | $I^-$ | iod*ide* |
| $Ca^{2+}$ | Calcium | $O^{2-}$ | ox*ide* |
| $Ba^{2+}$ | Barium | $S^{2-}$ | sulf*ide* |
| $La^{3+}$ | Lanthanum | $N^{3-}$ | nit*ride* |
| $Al^{3+}$ | Aluminum | $P^{3-}$ | phosph*ide* |

The following table gives examples of the formulas, the ions present and the names, of some ionic compounds.

It is important to note that when written as part of a chemical formula, the ions are not written with a charge, but when alone---like in aqueous solution---the charges are included.

# Inorganic Chemistry: Introduction to Coordination Chemistry

**Table 1.2** Examples of Simple Type I ionic compounds.

| Compound | Ions Present | Name |
|---|---|---|
| LiF | $Li^+$, $F^-$ | Lithium Fluoride |
| NaCl | $Na^+$, $Cl^-$ | Sodium Chloride |
| KBr | $K^+$, $Br^-$ | Potassium Bromide |
| RbI | $Rb^+$, $I^-$ | Rubidium Iodide |
| MgO | $Mg^{2+}$, $O^{2-}$ | Magnesium Oxide |
| CaS | $Ca^{2+}$, $S^{2-}$ | Calcium Sulfide |
| AlN | $Al^{3+}$, $N^{3-}$ | Aluminum Nitride |

Since all chemical compounds have to be electrically neutral (zero net charge) then combinations of cations and anions in ionic compounds must balance out their electrical charges. For any ionic compound,

Total Charge of Cations + Total Charge of Anions = Zero.

So, if we have an ionic compound, barium chloride, what is its correct formulation? We determine the formula of an ionic compound by adjusting the relative numbers (simplest whole number ratio) of the cations and anions so that electrical neutrality is achieved. To determine the correct formulation of barium chloride, one must locate the positions of barium and chlorine in the periodic table. Barium is found in Group 2, so that means it forms ions of a +2 charge.

So, since barium is a Group 2 metal and loses two electrons (barium → $Ba^{2+}$ + 2e$^-$), and as we have seen chlorine is a group 7 nonmetal and gains an electron (Chlorine + 1e$^-$ → $Cl^-$), then the ratio of $Ba^{2+}$ to $Cl^-$ must be 1:2, and therefore the formula of barium chloride is $BaCl_2$.

## Polyatomic Ions

The ions that have been discussed so far have been monatomic ions because they consist of only one atom bearing the electrical charge. There are groups of atoms that are linked together by covalent bonds which as a unit bear the electrical charge. These are called polyatomic ions. In a polyatomic ion the superscript is the electrical charge on the whole polyatomic ion unit. Many polyatomic ions are oxyanions, which have a central atom and peripheral oxygen atoms that make up a discrete unit.

Table 1.3 lists the common polyatomic ions that students should remember. It is important to note that there are very few polyatomic ions that are anionic. These are relegated to the cations formed from protonation of amines, such as ammonia, methylamine, dimethylamine, etc.

**Table 1.3** Common Polyatomic Ions

| Cations / Positive Charged Ions | | | |
|---|---|---|---|
| $NH_4^+$ | ammonium | $CH_3NH_3^+$ | methylamine |
| **Anions / Negative Charged Ions** | | | |
| $OH^-$ | hydroxide | $NO_2^-$ | nitrite |
| $CN^-$ | cyanide | $NO_3^-$ | nitrate |
| $CH_3COO^-$ | acetate | $OCl^-$ | hypochlorite |
| $MnO_4^-$ | permanganate | $ClO_3^-$ | chlorate |
| $HCO_3^-$ | bicarbonate | $ClO_4^-$ | perchlorate |
| $CO_3^{2-}$ | carbonate | $SO_3^{2-}$ | sulfite |
| $Cr_2O_7^{2-}$ | dichromate | $SO_4^{2-}$ | sulfate |
| $PO_4^{3-}$ | phosphate | $AsO_4^{3-}$ | arsenate |

**Type II ionic compounds**

The Type II ionic compounds all have transition metals or heavy main group metals as the cation. Most transition metals have the ability to exist in more than one oxidation state, and this can lead to confusion in nomenclature of their ionic compounds unless it is addressed in some way. For example the ionic compound generically named "iron oxide" could be FeO or $Fe_2O_3$. Another example is "copper chloride", which could either be CuCl or $CuCl_2$. The initial way to deal with this is now called the "older name", and is still used, particularly by chemical vendors, who sometimes use these older names. Some of these older names are listed in Table 1.4 along with comparisons of newer more systematic names.

The only way to differentiate between the compounds is to list the oxidation state of the transition metal ion, and that is done by using Roman numerals after the transition metal. Thus, FeO would be iron (II) oxide, and $Fe_2O_3$ would be named iron (III) oxide. Some of these systematic names are listed in Table 1.4.

**Table 1.4** Common Type (II) Cations

| Ion | Systematic Name | Older Name |
|---|---|---|
| $Cr^{+2}$ | chromium (II) | chromous |
| $Cr^{+3}$ | chromium (II) | chromic |
| $Fe^{+2}$ | iron (II) | ferrous |
| $Fe^{+3}$ | iron (III) | ferric |
| $Co^{+2}$ | cobalt (II) | cobaltous |
| $Co^{+3}$ | cobalt (III) | cobaltic |
| $Cu^{+}$ | copper (I) | cuprous |
| $Cu^{+2}$ | copper (II) | cupric |
| $Sn^{+2}$ | tin (II) | stannous |
| $Sn^{+4}$ | tin (IV) | stannic |
| $Pb^{+2}$ | lead (II) | plumbous |
| $Pb^{+4}$ | lead (IV) | plumbic |
| Au+ | gold (I) | aurous |
| $Au^{+3}$ | gold (III) | auric |

In the older name system, the common ion of the metal that has the lowest oxidation state has a name ending in *-ous*, whereas the one with the higher charge has a name ending in *-ic*. We will use the systematic name exclusively in this text. The best way to get acquainted or re-familiarize yourself with the formulation and the nomenclature of Type (I) and Type (II) ionic compounds is to work the problems at the end of the chapter.

## Main-Group compounds (Covalent Compounds)

Covalent compounds of the main group elements have a similar set of nomenclature rules as the ionic compounds discussed above, except for a striking difference. The elements have prefixes that are used to denote the numbers of atoms present in these molecules, but the prefix mono- is never used for the first element of the compound, but it is for the second element. Table 1.5 gives the prefixes used to indicate the numbers used in the chemical formulas of covalent compounds of the non-metals.

**Table 1.5** Prefixes for covalent compounds.

| Prefix | Number of atoms in the compound |
|---|---|
| *mono-* | 1 |
| *di-* | 2 |
| *tri-* | 3 |
| *tetra-* | 4 |
| *penta-* | 5 |
| *hexa-* | 6 |
| *hepta-* | 7 |
| *octa-* | 8 |
| *nona-* | 9 |
| *deca-* | 10 |

The covalent compounds containing oxygen often sound somehow wrong with many of the prefixes. For example monooxide is contracted to monoxide, and tetraoxide is contracted to tetroxide. So, contractions like these, for oxygen in particular, that make the name sound better are used to make pronunciation easier.

Many main-group compounds were discovered years ago and have common names that continue to be used: $H_2O$ is water, $NH_3$ is ammonia, $N_2H_4$ is hydrazine, $PH_3$ is phosphine, NO is nitric oxide, and $N_2O$ is nitrous oxide which is also known as laughing gas.

## Coordination Chemistry: The Early Years

Until mankind retreated from the non-rational philosophies of *mysticism* that characterized the Middle Ages, the pseudo-science of *alchemy* was doomed to failure. Eventually, the adoption of the *scientific method* allowed humanity to break off our shackles of non-objective irrationality and advance into the modern age of science using measurements and reason.

Currently, with the accomplishments and discoveries of over two centuries of hard science at our command, we are expanding the scope and dimension of scientific discoveries at an unbelievable rate. Driven by the thirst for knowledge and the prize of economic incentives, our global culture is set to accomplish incredible things in the years to come.

The history of coordination chemistry is a case study of this process. Alchemy set its sight on the goal of transmutation of the elements into nobler substances, such as gold, but the use of non-rational procedures, magic, and mysticism clouded the real advances in technology and discovery that were made by our predecessors. Now, with the recent advances in organometallics, nano-technology, and bioinorganic chemistry, the future has never looked brighter for the science of inorganic chemistry.

## The History of Coordination Compounds

Coordination compounds are typically characterized by four or six ligands (from the Latin word ***ligare***, which means "to bind") in a coordination sphere bonded to a metal atom or ion in a tetrahedral, square planar or octahedral geometry. This initial chapter starts an introductory investigation of coordination chemistry by putting its history into perspective and introducing some typical ligands and metal complexes.

The discovery of coordination compounds and the subsequent bonding explanations should be viewed against the larger history of progress in chemistry in understanding atomic structure of atoms, the periodic table of elements, and chemical bonding.

Coordination compounds were used for years before their true compositions and structures were determined. Heinrich **Diesbach** created *"Prussian Blue"* by accident in 1704, which is now formulated as $Fe_4[Fe(CN_6)]_3 \cdot H_2O$, by mixing contaminated potash with iron sulfate. Instead of red, which he was expecting, it was purple, then a deep blue when it further concentrated. This is the first, or one of the first, known coordination compounds. It could also be considered to be the first organometallic species synthesized.

**Joseph Proust** is best known in connection with a long controversy with **C. L. Berthollet** who was led by his doctrine of mass-action to deny that substances always combine in constant and definite proportions. Proust maintained that compounds always contain definite quantities of their elements. In 1799 he proved that carbonate of copper, whether natural or artificial, always has the same composition, and later he showed that the two oxides of tin and the two sulfides of iron always contain the same relative weights of their components and that no intermediate indeterminate compounds exist.

**Antoine Lavoisier** published *Reflexions sur le Phlogistique* (1783), where he showed the phlogiston theory to be inconsistent. In *Methods of Chemical Nomenclature* (1787), he invented the system of chemical nomenclature still largely in use today, including names such as sulfuric acid, sulfates, and sulfites.

His *Traité Élémentaire de Chimie (Elementary Treatise of Chemistry,* 1789) was the first modern chemical textbook, and presented a unified view of new theories of chemistry, contained a clear statement of the Law of Conservation of Mass, and denied the existence of phlogiston. In addition, it contained a list of elements, or substances that could not be broken down further, which included oxygen, nitrogen, hydrogen, phosphorus, mercury, zinc, and sulfur.

Contributions by Joseph Proust (***Law of Definite Proportions*** which is also called the Law of Constant Composition) and Antoine Lavoisier (father of modern chemistry - ***Law of Conservation of Mass***), along with many others, led John Dalton to formulate the first atomic theory.

**John Dalton** (1766-1844) developed the first useful ***atomic theory of matter*** around 1803. In his studies on meteorology, Dalton concluded that evaporated water exists in the air as an independent gas. Dalton reasoned that if water and air were composed of discrete individual particles, then evaporation might be viewed as a mixing of water particles with air particles. Dalton developed the hypothesis that the *sizes* of the particles making up different gases must be different while trying to explain the results of his experiments. In **1808** he published his *New System of Chemical Philosophy*.

Throughout the 1$^{st}$ half of the nineteenth century many crystalline samples of various cobalt ammonates were synthesized. These compounds are highly colored, and the names given to them (for example, roseo "red", luteo "deep yellow", and purpureo "purple" cobalt chlorides) reflected these colors. After the synthesis of these cobalt ammonates, a great deal of experimental work was performed by various chemists to try to determine what structure these beautifully colored compounds adopted.

**Dmitri Mendeleev** in **1869** published his first periodic table. The problems inherent in this wonderful contribution to chemistry have led to the discovery of new elements that he predicted, and the eventual rationalization and realization that the periodic table is associated with electronic structure.

In the 2$^{nd}$ half of the nineteenth century other ammonates of chromium and platinum were prepared. No theoretical basis was developed to account for the bonding and structure of these compounds however, and the chemistry languished because there was no rational basis for synthesis and structure of the new compounds.

In *1869-71 Christian Wilhelm Blomstrand* first proposed his *chain theory* to describe the structure of the cobalt ammonate chlorides and other series of transition metal ammonates. It was based upon a completely logical but incorrect analogy with the bonding in organic amines.

Blomstrand, who knew that the fixed valence (*oxidation state*) of cobalt was established at 3, came up with a structure where he linked together cobalt atoms, ammonia groups and chlorine atoms to produce a linear chain structure of $CoCl_3 \cdot 6NH_3$. Since it was based on the prevailing ideas of the time, this was considered to be a reasonable structure.

It was assumed that a metallic atom or ion could replace the hydrogen atoms of ammonia just as organic moieties do in the formation of amines such as methylamine $[CH_3NH_2]$, dimethylamine $[(CH_3)_2NH]$, and trimethylamine $[(CH_3)_3N]$. For example, the existence of two forms of $PtCl_4(NH_3)_2$ was rationalized by the two formulas:

$Cl_3Pt—NH_3—NH_3Cl$ and $Cl_2Pt(NH_3Cl)_2$.

**Svante Arrhenius** was a Swedish chemist best known for his theory that electrolytes, (which are substances that dissolve in water to yield a solution that conducts electricity) are separated or dissociated into electrically charged particles, or ions, even when there is no current flowing through the solution.

This resulted in his thesis (**1884**) "*Recherches sur la conductibilité galvanique des électrolytes*" (Investigations on the Galvanic Conductivity of Electrolytes). Arrhenius concluded that when electrolytes are dissolved in water they become to varying degrees split or dissociated into electrically opposite positive and negative ions (cations and anions).

In 1903 he was awarded the Nobel Prize for Chemistry.

S.M. **Jörgensen** (1837-1914) extended Blomstrand's chain theory in **1884**. Before Jörgensen began his extensive studies on the synthesis of "complex" metal compounds it was known that the reaction of metal halides and other salts with neutral molecules could give stable compounds. Many of these compounds could easily be formed in aqueous solutions.

The chain theory is of interest mainly because it became a matter of contentious debate between Jörgensen and Werner, and the study of this heated discussion gives insight into the workings of chemistry and science. The contentious debate between the two rivals (Jörgensen was older, Werner was younger) provided the stimulus for further research.

As more experimental data became available, the chain theory required increasing modification. Ultimately this patch-work theory was finally discarded, but it had a long period of popularity despite all its flaws. Discussion has been made about the possible damage that was done to science by this strong belief in an incorrect model, but it was an important part of the process of discovery.

The contentious scientific debate is described in the following passages. Jörgensen, who was a student of Blomstrand, proposed some amendments to Blomstrand's depiction of the cobalt ammonates.

In the end, Alfred Werner's theories were found to be essentially correct, as discussed in the next few pages, but the scientific debate on the theory and structure of metal complexes has been extremely valuable to our discipline of chemistry. It sparked a growing interest in metal complexes that has not stopped, and has led to many very important discoveries in materials science and medicine.

Structures of the cobalt ammonate chlorides as depicted by Blomstrand and Jörgensen are shown in Figure 1.3.

**Figure 1.3** The Blomstrand-Jörgensen Chain Theory

(a) Blomstrand's structure of $CoCl_3 \cdot 6NH_3$

(b) Jörgensen's structures of the other four members of the cobalt ammonate series plus the iridium substituted compound

(a) $CoCl_3 \cdot 6NH_3$

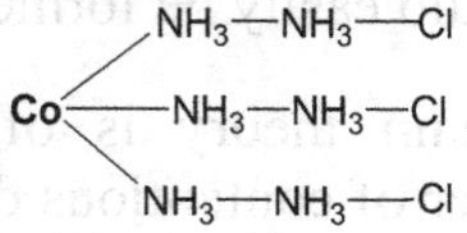

$$Co\begin{cases} NH_3-NH_3-Cl \\ NH_3-NH_3-Cl \\ NH_3-NH_3-Cl \end{cases}$$

(b) Series

(1) $CoCl_3 \cdot 6NH_3$

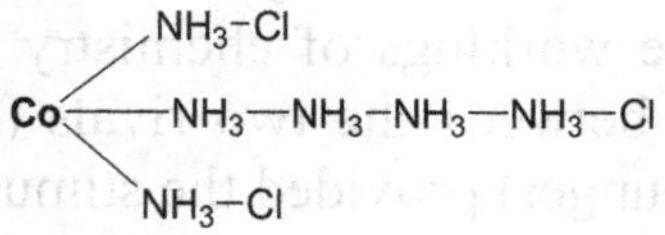

$$Co\begin{cases} NH_3-Cl \\ NH_3-NH_3-NH_3-NH_3-Cl \\ NH_3-Cl \end{cases}$$

(2) $CoCl_3 \cdot 5NH_3$

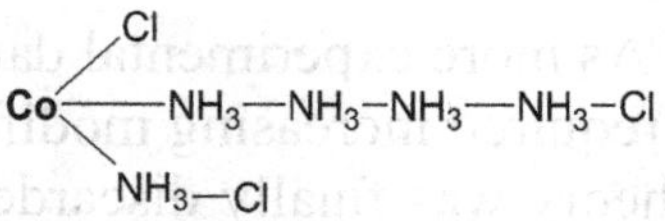

$$Co\begin{cases} Cl \\ NH_3-NH_3-NH_3-NH_3-Cl \\ NH_3-Cl \end{cases}$$

(3) $CoCl_3 \cdot 4NH_3$

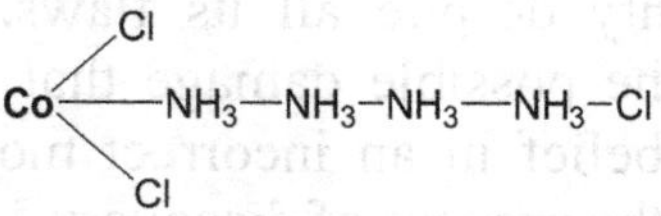

$$Co\begin{cases} Cl \\ NH_3-NH_3-NH_3-NH_3-Cl \\ Cl \end{cases}$$

(4) $IrCl_3 \cdot 3NH_3$

$$Ir\begin{cases} Cl \\ NH_3-NH_3-NH_3-Cl \\ Cl \end{cases}$$

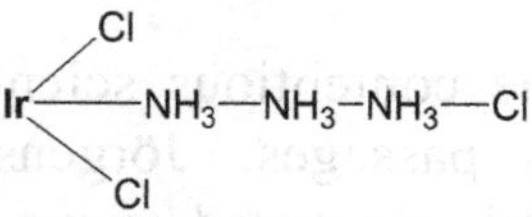

First, Jörgensen proposed that these complex compounds were monomeric, i.e. they only contained one metal.

In their laboratories, the number of chloride ions precipitated as silver chloride (see Table 1.) was determined by the addition of aqueous silver nitrate, as represented in the following equation:

$$AgNO_{3\,(aq)} + Cl^-_{(aq)} \rightarrow AgCl_{(s)} + NO_3^-_{(aq)}$$

Next, to account for the rates at which various chlorides were precipitated he adjusted the distance of these chloride groups away from the cobalt.

Since the first chloride is precipitated much more rapidly by silver nitrate than the other chlorides, it was placed farther down the chain and under less influence of the cobalt atom. Jörgensen assumed that if the chloride was directly attached to the cobalt, then it would not be available to participate in the precipitation reaction with silver nitrate.

The proposed structures for the first three cobalt ammonate chlorides are shown in Figure 1, and it can be seen in the structure of the second compound in the (b) series that one chloride is directly attached to the cobalt and is therefore unavailable to be precipitated by silver nitrate. In the third compound, two chlorides are similarly pictured, and are thus unavailable for precipitation by silver nitrate.

These changes improved the viability of the chain theory, but a number of unanswered questions remained. For example, why are there only 6 ammonia molecules in the compounds and not 8 or 4? Why do different ammonia molecules that are unique depending on their positions in the chain, not react differently? Even though these questions remained unanswered, the Blomstrand-Jörgensen theory of the cobalt ammonates was accepted anyway.

Given the success of organic chemists in describing the structural units and fixed atomic valences found in carbon-based compounds, it was natural that these ideas be applied to metal ammonates. However, the compounds did not behave as did most organic compounds; many were soluble in water for example, yet they did not behave as did typical inorganic slats such as sodium chloride, which is colorless. The favored way to write the chemical formulas for these substances indicated that they consisted of ionic salts and molecular-like substituents.

This lack of knowledge about the true nature of the bonding was reflected in the dot used in the formula to connect $CoCl_3$ to the correct number of ammonias (for example, $CoCl_3 \cdot 6NH_3$).

**Table 1.6:** The Cobalt Ammonate Chlorides
(Data Available to Blomstrand, Jörgensen, and Werner)

| Formula | Conductivity | # of $Cl^-$ ions precipitated | Electrolytic Behavior |
|---|---|---|---|
| NaCl | | 1 | 1:1 electrolyte |
| $CaCl_2$ | | 2 | 1:2 electrolyte |
| $LaCl_3$ | | 3 | 1:3 electrolyte |
| $CoCl_3 \cdot 6NH_3$ | High | 3 | 1:3 electrolyte |
| $CoCl_3 \cdot 5NH_3$ | Medium | 2 | 1:2 electrolyte |
| $CoCl_3 \cdot 4NH_3$ | Low | 1 | 1:1 electrolyte |
| $CoCl_3 \cdot 4NH_3$ | Low | 1 | 1:1 electrolyte |
| $IrCl_3 \cdot 3NH_3$ | Zero | 0 | nonelectrolyte |

The $CoCl_3 \cdot 3NH_3$ compound wasn't known at the time. Jörgensen did, however, manage to prepare, after considerable time and effort, the analogous iridium ammonate chloride, but since it was found to have *no ionizable chlorides* then the only conclusion possible was that it must be a neutral compound.

The Blomstrand-Jörgensen chain theory was in deep trouble due to the research efforts of one of its investigators!

**Werner and Coordination Theory**

In **1891, Alfred Werner**, at age twenty-five, published a paper titled "Contribution to the Theory of Affinity and Valence," where he rejected the usual concepts of valence and affinity, (attraction) of atoms for each other.

In **1892**, the following year, Werner proposed **coordination theory**, which formed the basis for a new field of inorganic chemistry. Werner observed the difficulties that inorganic chemists were having in explaining the bonding in metal "complexes", and he was aware that the established ideas of organic chemistry seemed to lead only into more confusion.

The recognition of the true nature of metal "complexes" was set out in his classic work *Neuere Anschauungen auf dem Gebiete der anorganischen Chemie*

(Newer Ideas in Inorganic Chemistry) in 1905; he received the Nobel Prize for this work in 1913.

Accounts say that, "According to his own statement, the inspiration in 1892 came to him like a flash. One morning at two o'clock he awoke with a start: the long-sought solution of this problem had lodged in his brain. He arose from his bed and by five o'clock in the afternoon the essential points of the coordination theory were achieved."

In developing his theory Werner made a major departure from prevailing theory.

Werner discarded Blomstrand's and Jorgensen's chain theory in favor of a centralized construction for metal complexes about which other atoms arranged or coordinated themselves around.

The primary valence, or ionizable valence, corresponded to what we call today the *oxidation state*; for cobalt, it is the $3^+$ state. The secondary valence is more commonly called *the coordination number* (the number of atoms or groups directly attached to an atom). For cobalt it is six.

Thus, the primary valence is the oxidation state, and the secondary valence is the coordination number.

He found that a few coordination numbers are prevalent, namely two, four, six, and eight, with six by far the most common. Werner proposed likely spatial configurations for those coordination numbers and explained the existence of geometrical isomers.

He predicted the existence of optical isomers, which he eventually isolated, for some cobalt compounds of coordination number 6; these were the first known optically active inorganic compounds.

It is interesting to see how Werner used his postulates to explain the properties of the cobalt (III) chloride-ammonia complexes listed in Table 1.

Thus, the compound $CoCl_3 \cdot 6NH_3$ was represented as $[Co(NH_3)_6]Cl_3$ since there are three readily available chloride ions.

In solution there were therefore four ions, $[Co(NH_3)_6]^{3+}$ and the three chloride ions. This is an important distinction!

The conductivity data (see Table 1.) could then be explained because $CoCl_3 \cdot 6NH_3$ was actually $[Co(NH_3)_6]Cl_3$, and the following dissolution process occurs in aqueous solution:

$$[Co(NH_3)_6]Cl_3 \rightarrow [Co(NH_3)_6]^{3+} + 3\ Cl^-$$

It is important for you the student to be able to recognize that some atoms, molecules or ions are bound directly to the metal atom in a "covalent" type fashion, which will be discussed later, and that counter-ions may be present as well which are ionically bonded to the metal complex, just as $Na^+$ and $Cl^-$ are bonded together in common table salt. In the case of the $[Co(NH_3)_6]Cl_3$ complex, the formula indicates that the ammonia molecules are bound to the cobalt atom *covalently* for lack of a better term at this point, and that the three chloride ions are just counter-ions.

The formulas for the other cobalt ammonate compounds can be determined in a similar way. Two compounds, a violet and a green form of $CoCl_3 \cdot 4NH_3$, were puzzling to chemists of the time because they have the same formula although they are obviously different chemical compounds.

Werner concluded that these two compounds were geometrical isomers and were the *cis* and *trans* forms, where coordinated atoms are at the corners of an octahedron.

As might be expected, Werner's radical departure from long-accepted chemistry theories and bonding concepts met with disapproval and controversy.

Jörgensen criticized Werner's theory on several points: (1) the postulate that all coordinating groups occupy equivalent positions in the coordination sphere, (2) Werner's designation of geometrical isomers of such compounds as $[CoCl_2(NH_3)_4]Cl$, and (3) the prediction of metal compounds that were then unknown.

These criticisms, and others, caused Werner and his associates to diligently produce experimental evidence to support his coordination theory in detail.

The coordination theory that Werner proposed has not been discredited by further research over time, but has served as a framework and a guide for all coordination chemists.

Werner next turned to the geometry of the secondary valence (or coordination number). As shown in Table 2, six ammonia molecules placed about a central metal atom or ion might adopt one of several different common geometries, including **hexagonal planar**, **trigonal prismatic**, and ***octahedral***. The table compares some information about the predicted and actual number of isomers for a variety of substituted coordination compounds, in particular the cobalt ammonates.

There is a need to make a few comments about the information in this table before discussing the data. Note that the symbols for the compounds use M for the central metal atom and A's and B's for the ligands. ***Isomers* are defined here as compounds that have the same numbers and types of chemical bonds but differ in the spatial arrangements of those bonds**. The number of predicted isomers refers to the number of arrangements in space that are theoretically possible for the given geometry.

For the planar MA$_5$B case, for example, there is only one geometry possible, even though there are numerous ways to draw it. Draw these yourself, and make models!!

**Table 1.7.** The Number of Actual versus Predicted Isomers of Three Possible Geometries of Coordination Number 6

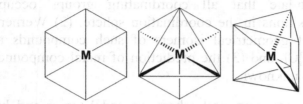

| Formula | Planar | Octahedral | Trigonal Prism | Observed |
|---|---|---|---|---|
| MA$_5$B | 1 | 1 | 1 | 1 |
| MA$_4$B$_2$ | 3 | 2 | 3 | 2 |
| MA$_3$B$_3$ | 3 | 2 | 3 | 2 |

In comparing different isomers of the same geometry it does not matter which position is occupied by the one B ligand since all the positions are equivalent; thus, there is only one isomer for the MA$_5$B case.

Now, one can inspect the data in Table 2. In the MA$_5$B case, only one isomer of the cobalt ammonates could actually be synthesized. This result is consistent with all three of the proposed geometries, and doesn't help in determining which geometry is favored by the cobalt ammonate coordination compounds.

However, for the MA$_4$B$_2$ case, Werner could prepare only *two* isomers! Importantly, this actual number of isomers synthesized matched the possible number of isomers predicted for the octahedral geometry. Also, and just as importantly, for the hexagonal planar and trigonal prism cases there were *three* possible predicted isomers for each of those geometries.

Assuming that Werner hadn't missed the preparation of an isomer, this indicated that the coordination geometry for six ligands around cobalt is octahedral.

For the $MA_3B_3$ case only the octahedral geometrical configuration predicts the same number of isomers that were actually synthesized by Werner. By analyzing a large number of series of coordination compounds of different metals, Werner could predict that two isomers would be found for the $CoCl_3 \cdot 4NH_3$ case. These two isomers proved to be difficult to prepare, but in 1907 Werner was finally successful in their synthesis.

Werner synthesized two isomers of $CoCl_3 \cdot 4NH_3$, one a striking violet and the other a bright green color.

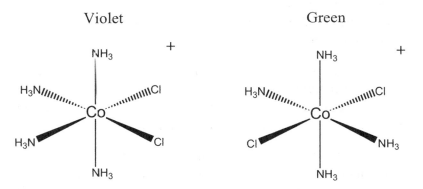

Although all this would not be considered conclusive proof but as "negative" evidence by some philosophers of science---since it was the *absence* of an isomer that constituted the evidence---the case for Werner's coordination theory grew stronger.

However, this negative proof was enough for Jörgensen. In 1907 he dropped his opposition to Werner's "preposterous" coordination theory.

## Inorganic Chemistry: Introduction to Coordination Chemistry

The march of coordination chemistry continued with new discoveries, which forced a rethinking of older organic theories, and the emergence of newer bonding theories.

Once the electron was discovered in the early part of the twentieth century (J.J. **Thomson**, Robert **Millikan**), then **G.N. Lewis** was able to explain some aspects of bonding on the basis of his electron-dot formulas and the octet rule, which we will cover in-depth in Chapter 2.

G. N. Lewis was trying to explain valence to his students in **1902** when he depicted atoms as a concentric series of cubes with electrons at each corner. This "cubic atom" explained the eight groups in the periodic table and presented his theory that chemical bonds are formed by electron sharing to give each atom a complete set of eight, an octet. These early *Lewis Electron Dot Structures* would later be modified in another classroom discussion.

In **1923 Lewis** redefined acids as any atom or molecule with an incomplete "octet" that were thus capable of accepting electrons from another atom; bases were, of course, electron donors. Thus began the useful *Lewis Acid/Base Theory*.

In his book, "The Electronic Theory of Valency", which was published in **1927** as a culmination of many years' interest in the nature of covalent and dative bonds, **Nevil Sidgwick** discussed the electronic interpretation of coordination chemistry and discussed the covalent nature of the chemical bond between metal and ligand, since Werner had shown that it wasn't ionic.

In **1929, Hans Bethe** used J. Becquerel's basic idea (1928) of a crystal field and formulated an exact theory he called *"Crystal Field Theory"* which used symmetry concepts to describe the electronic levels of gaseous metal ions.

In **1931-32, Linus Pauling** used the ideas of Sidgwick and Lewis to discuss the covalent bonding between a metal and a ligand in quantum mechanical language, and used metal orbitals to construct a hypothetical "hybrid orbital" which the metal could use to bond to a ligand in a covalent fashion. This hybrid orbital was described as a $d^2sp^3$ hybrid orbital. Linus Pauling's valence-bond theory will be discussed in Chapter 2 and Chapter 5.

**J.H. Van Vleck** made the first real application of Bethe's Crystal Field Theory to the chemistry of the transition metals in **1932**. He succeeded in explaining why the paramagnetism of some of the complexes of the first transition metal series existed using the Crystal Field Theory.

The valence-shell electron-pair repulsion (VSEPR), valence-bond (VB), and molecular orbital (MO) theories were fleshed out in the 1930s. We will cover VSEPR theory (by Sidgwick) in Chapter 2, and discuss valence-bond theory as set out by Linus Pauling. Molecular orbital theory will be discussed later in the text, in Chapter 6.

Crystal Field Theory was largely ignored before WW II, and then utilized heavily in the 1950's once it was "discovered" in the literature.

This marks the beginning of the modern age of coordination chemistry.

## Older formulas vs. Modern formulas

The formulation of metal complexes can now no longer be expressed in the older form such as listed in Table1 since that form does not reveal any structural information. Since two different metal complexes can have the same formula ($CoCl_3 \cdot 4NH_3$ as shown above), then it is necessary to formulate them differently. The formulations must be systematic and describe covalent bonding and the ionic nature of the metal complex as accurately as possible.

This is taken up in the next section.

## Inorganic Chemistry: Introduction to Coordination Chemistry

**Anatomy of a metal complex (Modern Formulas)**

A transition metal complex has several parts as can be seen in the following scheme: (1) a central metal atom, (2) ligands, and (3) it may possess a counter ion such as an anion.

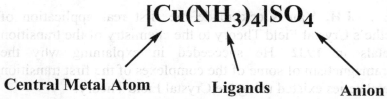

The ligands are bound directly to the central metal atom, and are said to be in the primary coordination sphere. The counter-ions, such as the sulfate anion in the above copper complex, are bound to the metal complex by electrostatic attraction, and are said to be in the secondary coordination sphere. The above metal complex may dissolve in water, like a normal salt, to produce a cation and an anion:

$$[Cu(NH_3)_4]SO_4 \xrightarrow{water} [Cu(NH_3)_4]^{2+} + SO_4^{2-}$$

Alternatively, and more rarely, a metal complex may actually be an anion instead of a cation. An example is shown below:

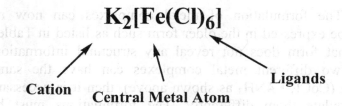

In this case, the iron complex is an anion, and it needs two potassium cations to balance its charge. It may dissolve in water to produce:

$$K_2[Fe(Cl)_6] \xrightarrow{water} 2\,K^+ + [Fe(Cl)_6]^{2-}$$

## Periodic table and trends

The periodic table is divided into three types of elements: (1) *metals*, (2) *semimetals* or *metalloids*, and (3) *nonmetals*. The metals make up the bulk of the periodic table and the chemistry of the transition metals is the focus of this text. The nonmetals are also of importance to the chemistry of the transition metals, since they provide the ligands that react with the metals.

Metals are distinguished chemically from the nonmetals by the absence of negative oxidation states. (There are exceptions in organometallic compounds) Metals have the ability to form positive ions, ionic bonds, and basic oxides.

The vertical columns or *Groups* of the periodic table contain elements having similar chemical and physical properties, and several groups of elements have distinct names that you must know.

Elements in Group I are known as the *alkali metals* (except for hydrogen). The word alkali comes from the ancient Arabic word *al-qali*, which is the word for plants whose ashes give water solutions that were slippery and burned the skin. The alkali metals are always found as +1 charged cations.

Similarly, the Group II metals are called the *alkaline earth metals*, since the metal hydroxides form basic solutions and the Group II cations, which are always +2 charged, usually form insoluble salts.

The two rows at the bottom of the periodic table, the so-called *f-block elements*, are the *lanthanides* and the *actinides*, which are also called the *inner transition metals*. The f-block elements have a common +3 oxidation state.

Group VII elements are called the *halogens*. Group VI elements are called the *chalcogens* from the Greek word, *khalkos*, for copper, whose ores often contain them. These are nonmetals.

The ***transition metals*** are the so-called ***d-block elements*** (which follow Group II elements) that fill the fourth, fifth, and sixth periods in the center of the periodic table. The transition metals have varying oxidation states, with many different oxidation states each, with the most common oxidation states being +2, and +3. Transition metals can use electrons from more than one subshell, so they exhibit multiple oxidation states. As a general rule, the higher the oxidation state of the metal, the more covalent is the bonding.

Chemists often use the term *transition metal* for elements that have partially filled d-orbitals. Thus, Zn, Cd, and Hg are often excluded from the list, as well as Sc, Y, and La, and it is no surprise that they do exhibit chemistry that resembles the representative main group elements.

Transition metals and their compounds or complexes have characteristic properties such as:
- *variable oxidation states,*
- *small, compact atoms,* which decrease across a period and then increases towards the end,
- *stable compounds* of varying covalency in far more numerous examples, geometries and types than do the metals from the main group,
- *magnetic properties* such as paramagnetism and diamagnetism,
- *highly colored complexes* that span the rainbow (names of some of the transition metals express this fact---iridium, rhodium, chromium, etc.),
- *catalytic activity* such as with enzymes and industrially used catalysts and,
- *variable coordination geometries* such as square planar complexes, which are unknown for the main group metals. It is the study of these fascinating properties of the transition metals and their coordination chemistry that is the focus of this book.

## Electronic Configurations

On pages 35 and 36 are listed the *electronic configurations* of the first 86 elements, and the atomic radii of selected elements, which include the first three rows of the transition metals. The transition metals are listed in bold.

Electron configurations are extremely important to inorganic chemists and help us to understand the reactivity of transition metal ions. The loss of certain numbers of electrons from metal atoms to form metal ions can be understood much better once the electron configuration information is available.

For example, scandium is nearly always found in the $Sc^{+3}$ oxidation state. This makes sense because the electron configuration of Sc metal is: $Sc = [Ar]\ 4s^2\ 3d^1$. If Sc metal loses three electrons then the electron configuration of $Sc^{+3}$ would be that of Ar.

Since most of the transition metals have $ns^2$ (where n=3, 4 or 5) as part of their electron configuration, then it is no surprise that the +2 oxidation state for transition metals is so very common. In transition metal complexes it is assumed by convention that the $ns^2$ electrons are always the first to go when the metal is oxidized. Also, as it will be restated in the section on electron counting in Chapter 7., the electron configuration given here is only for isolated atoms of that element (gaseous atoms). If ligands are around a metal in some metal complex, then the electronic configurations noted here are not considered to be valid.

## Atomic Radii of the Transition Metals

Radii of the transition metals are determined in the solid state by using X-Ray diffraction to measure the metal-metal distances. The radii of the transition metals vary over a narrow range in each period, which is determined by the outermost valence electrons, which are in the 4s, 5s or 6s orbitals for each of the three periods respectively.

**A graph of atomic radii of the 3d, 4d and 5d transition metal series.**

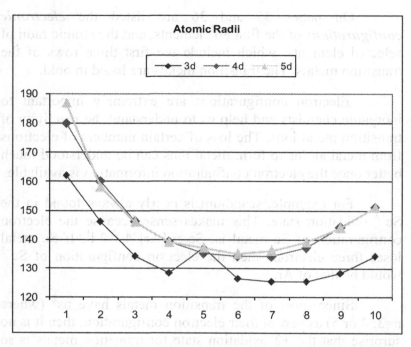

The radii of the transition metals tends to decrease in size as we go across the period. This is due to the effective nuclear charge increasing across a period, and the fact that d-electrons are not very good at shielding other electrons in the valence shell from the nuclear charge. An interesting observation is that the radii of the $2^{nd}$ and $3^{rd}$ row transition metals (the fifth and sixth period) are almost identical. The reason for this is that the inner transition elements, the lanthanide elements, are inserted into the Periodic Table between La and Hf. The filling of the 4f orbitals is accompanied by a steady decrease in the atomic radii to the point that the $3^{rd}$ row transition elements have acquired a size almost identical with those in the $2^{nd}$ row. This significant effect is called the ***lanthanide contraction***.

(Interestingly, both Lanthanum and Lutetium have a $5d^1$ configuration and there is controversy as to which is the first transition metal in the third row of the transition metals!)

## Table 1.8 Electron Configurations of the First 86 Elements

| Z | Element | Electron Configuration |
|---|---------|------------------------|
| 1 | H | $1s^1$ |
| 2 | He | $1s^2$ |

| Z | Element | Electron Configuration |
|---|---------|------------------------|
| 3 | Li | [He] $2s^1$ |
| 4 | Be | [He] $2s^2$ |
| 5 | B | [He] $2s^2\, 2p^1$ |
| 6 | C | [He] $2s^2\, 2p^2$ |
| 7 | N | [He] $2s^2\, 2p^3$ |
| 8 | O | [He] $2s^2\, 2p^4$ |
| 9 | F | [He] $2s^2\, 2p^5$ |
| 10 | Ne | [He] $2s^2\, 2p^6$ |

| Z | Element | Electron Configuration |
|---|---------|------------------------|
| 11 | Na | [Ne] $3s^1$ |
| 12 | Mg | [Ne] $3s^2$ |
| 13 | Al | [Ne] $3s^2\, 3p^1$ |
| 14 | Si | [Ne] $3s^2\, 3p^2$ |
| 15 | P | [Ne] $3s^2\, 3p^3$ |
| 16 | S | [Ne] $3s^2\, 3p^4$ |
| 17 | Cl | [Ne] $3s^2\, 3p^5$ |
| 18 | Ar | [Ne] $3s^2\, 3p^6$ |

| Z | Element | Electron Configuration | Atomic Radii |
|---|---------|------------------------|--------------|
| 19 | K | [Ar] $4s^1$ | 230 |
| 20 | Ca | [Ar] $4s^2$ | 197 |
| 21 | Sc | [Ar] $4s^2\, 3d^1$ | 162 |
| 22 | Ti | [Ar] $4s^2\, 3d^2$ | 146 |
| 23 | V | [Ar] $4s^2\, 3d^3$ | 134 |
| 24 | Cr | [Ar] $4s^1\, 3d^5$ | 128 |
| 25 | Mn | [Ar] $4s^2\, 3d^5$ | 137 |
| 26 | Fe | [Ar] $4s^2\, 3d^6$ | 126 |
| 27 | Co | [Ar] $4s^2\, 3d^7$ | 125 |
| 28 | Ni | [Ar] $4s^2\, 3d^8$ | 125 |
| 29 | Cu | [Ar] $4s^1\, 3d^{10}$ | 128 |
| 30 | Zn | [Ar] $4s^2\, 3d^{10}$ | 134 |
| 31 | Ga | [Ar] $4s^2\, 3d^{10}\, 4p^1$ | 135 |
| 32 | Ge | [Ar] $4s^2\, 3d^{10}\, 4p^2$ | |
| 33 | As | [Ar] $4s^2\, 3d^{10}\, 4p^3$ | |
| 34 | Se | [Ar] $4s^2\, 3d^{10}\, 4p^4$ | |
| 35 | Br | [Ar] $4s^2\, 3d^{10}\, 4p^5$ | |
| 36 | Kr | [Ar] $4s^2\, 3d^{10}\, 4p^6$ | |

| Z | Element | Electron Configuration | Atomic Radii |
|---|---------|------------------------|--------------|
| 37 | Rb | [Kr] $5s^1$ | 247 |
| 38 | Sr | [Kr] $5s^2$ | 215 |
| 39 | Y | [Kr] $5s^2\, 4d^1$ | 180 |

## Inorganic Chemistry: Introduction to Coordination Chemistry

| | | | |
|---|---|---|---|
| 40 | Zr | [Kr] $5s^2\ 4d^2$ | 160 |
| 41 | Nb | [Kr] $5s^1\ 4d^4$ | 146 |
| 42 | Mo | [Kr] $5s^1\ 4d^5$ | 139 |
| 43 | Tc | [Kr] $5s^1\ 4d^6$ | 135 |
| 44 | Ru | [Kr] $5s^1\ 4d^7$ | 134 |
| 45 | Rh | [Kr] $5s^1\ 4d^8$ | 134 |
| 46 | Pd | [Kr] $5s^0\ 4d^{10}$ | 137 |
| 47 | Ag | [Kr] $5s^1\ 4d^{10}$ | 144 |
| 48 | Cd | [Kr] $5s^2\ 4d^{10}$ | 151 |
| 49 | In | [Kr] $5s^2\ 4d^{10}\ 5p^1$ | 167 |
| 50 | Sn | [Kr] $5s^2\ 4d^{10}\ 5p^2$ | 154 |
| 51 | Sb | [Kr] $5s^2\ 4d^{10}\ 5p^3$ | |
| 52 | Te | [Kr] $5s^2\ 4d^{10}\ 5p^4$ | |
| 53 | I  | [Kr] $5s^2\ 4d^{10}\ 5p^5$ | |
| 54 | Xe | [Kr] $5s^2\ 4d^{10}\ 5p^6$ | |

| Z | Element | Electron Configuration | Atomic Radii |
|---|---|---|---|
| 55 | Cs | [Xe] $6s^1$ | 267 |
| 56 | Ba | [Xe] $6s^2$ | 222 |
| **57** | **La** | **[Xe] $6s^2\ 5d^1$** | **187** |
| 58 | Ce | [Xe] $6s^2\ 5d^0\ 4f^2$ | 182 |
| 59 | Pr | [Xe] $6s^2\ 5d^0\ 4f^3$ | 182 |
| 60 | Nd | [Xe] $6s^2\ 5d^0\ 4f^4$ | 182 |
| 61 | Pm | [Xe] $6s^2\ 5d^0\ 4f^5$ | 181 |
| 62 | Sm | [Xe] $6s^2\ 5d^0\ 4f^6$ | 180 |
| 63 | Eu | [Xe] $6s^2\ 5d^0\ 4f^7$ | 204 |
| 64 | Gd | [Xe] $6s^2\ 5d^0\ 4f^8$ | 179 |
| 65 | Tb | [Xe] $6s^2\ 5d^0\ 4f^9$ | 178 |
| 66 | Dy | [Xe] $6s^2\ 5d^0\ 4f^{10}$ | 177 |
| 67 | Ho | [Xe] $6s^2\ 5d^0\ 4f^{11}$ | 176 |
| 68 | Er | [Xe] $6s^2\ 5d^0\ 4f^{12}$ | 175 |
| 69 | Tm | [Xe] $6s^2\ 5d^0\ 4f^{13}$ | 174 |
| 70 | Yb | [Xe] $6s^2\ 5d^0\ 4f^{14}$ | 193 |
| 71 | Lu | [Xe] $6s^2\ 5d^1\ 4f^{14}$ | 174 |
| **72** | **Hf** | **[Xe] $6s^2\ 5d^2\ 4f^{14}$** | **158** |
| **73** | **Ta** | **[Xe] $6s^2\ 5d^3\ 4f^{14}$** | **146** |
| **74** | **W**  | **[Xe] $6s^2\ 5d^4\ 4f^{14}$** | **139** |
| **75** | **Re** | **[Xe] $6s^2\ 5d^5\ 4f^{14}$** | **137** |
| **76** | **Os** | **[Xe] $6s^2\ 5d^6\ 4f^{14}$** | **135** |
| **77** | **Ir** | **[Xe] $6s^2\ 5d^7\ 4f^{14}$** | **136** |
| **78** | **Pt** | **[Xe] $6s^1\ 5d^9\ 4f^{14}$** | **139** |
| **79** | **Au** | **[Xe] $6s^1\ 5d^{10}\ 4f^{14}$** | **144** |
| **80** | **Hg** | **[Xe] $6s^2\ 5d^{10}\ 4f^{14}$** | **151** |
| 81 | Tl | [Xe] $6s^2\ 5d^{10}\ 4f^{14}\ 6p^1$ | 171 |
| 82 | Pb | [Xe] $6s^2\ 5d^{10}\ 4f^{14}\ 6p^2$ | 175 |
| 83 | Bi | [Xe] $6s^2\ 5d^{10}\ 4f^{14}\ 6p^3$ | |
| 84 | Po | [Xe] $6s^2\ 5d^{10}\ 4f^{14}\ 6p^4$ | |
| 85 | At | [Xe] $6s^2\ 5d^{10}\ 4f^{14}\ 6p^5$ | |
| 86 | Rn | [Xe] $6s^2\ 5d^{10}\ 4f^{14}\ 6p^6$ | |

# Inorganic Chemistry: Introduction to Coordination Chemistry

## Terms and Definitions Chapter 1

*Alfred Werner* ---
*Christian Blomstrand* ---
*S.M. Jörgensen* ---
*Chain Theory* ---

*Ionizable Chlorides* ---

*Oxidation State* ---

*Coordination Number* ---

*Isomers* ---

*Primary and Secondary Valence* ---

*Metals* ---
*Semimetals* or *Metalloids* ---
*Nonmetals* ---

*Groups* ---
*Alkali Metals* ---
*Alkaline Earth Metals* ---
*Halogens* ---
*Chalcogens* ---

*Transition Metals* ---
*Inner Transition Metals* ---
*d-Block Elements* ---
*f-Block Elements* ---
*Lanthanides* ---
*lanthanide contraction* ---
*Actinides* ---
*Electronic Configurations* ---

# Inorganic Chemistry: Introduction to Coordination Chemistry

## Chapter 1 Problems and Exercises

1. Name the following Type I ionic compounds:
(a) $MgBr_2$  (b) $CaO$  (c) $Na_2S$  (d) $KNO_3$  (e) $RbBr$  (f) $LiF$
(g) $ScCl_3$  (h) $ZnS$  (i) $AlBr_3$  (j) $CdI_2$  (k) $SrSO_4$  (l) $Mg_3P_2$

2. Give the chemical formulas for the following Type I ionic compounds: (a) magnesium chloride $MgCl_2$ (b) sodium oxide $Na_2O$
(c) potassium sulfide $K_2S$ (d) rubidium arsenate $Rb_3(AsO_4)$ (e) calcium carbonate $CaCO_3$ (f) cesium sulfate $Cs_2SO_4$ (g) aluminum oxide $Al_2O_3$
(h) barium sulfate $BaSO_4$ (i) strontium nitrate $Sr(NO_3)_2$ (j) lithium phosphide $Li_3PO_4$
(k) aluminum hydroxide $Al(OH)_3$ (l) potassium permanganate $KMnO_4$
(m) zinc oxide $ZnO$ (n) scandium sulfite $Sc_2(SO_3)_3$ (o) cesium cyanide $CsCN$
(p) sodium bicarbonate $Na_2CO_3$ (q) potassium dichromate $K_2CrO_4$
(r) sodium nitrite $Na_2NO_2$ (s) ammonium chloride $NH_4Cl$ (t) sodium nitrate $NaNO_3$
(u) ammonium acetate $NH_4C_2H_3O_2$ (v) ammonium sulfate $(NH_4)_2SO_4$
(w) ammonium phosphate $(NH_4)_3PO_4$ (x) magnesium nitride $Mg_3N_2$
(y) sodium hypochlorite $NaClO_4$ (z) calcium chlorate $Ca_2ClO_3$

3. Name the following Type II ionic compounds:
(a) $SnF_2$  (b) $Co_2O_3$  (c) $SnF_4$  (d) $FeO_2$  (e) $Fe(ClO_4)_3$
(f) $Cr_2(SO_4)_3$  (g) $Ir(NO_3)_3$  (h) $MoO_3$  (i) $HgO$  (j) $MnO_2$
(k) $Pb(NO_3)_4$  (l) $CoBr_2$  (m) $TiCl_4$  (n) $Cu(NO_2)_2$
(o) $CuCN$  (p) $Cr_2(CO_3)_3$  (q) $PdCl_2$  (r) $Pb(HCO_3)_2$

4. Give the chemical formulas for the following Type II ionic compounds: (a) iron(II) chloride $FeCl_2$ (b) cobalt(III) oxide $Co_2O_3$
(c) chromium(III) oxide $Cr_2O_3$ (d) iron(III) hydroxide $Fe(OH)_3$
(e) titanium(IV) bromide $TiBr_4$ (f) copper(II) sulfate $CuSO_4$
(g) chromium(III) cyanide $Cr(CN)_3$ (h) rhodium(III) chloride $RhCl_3$
(i) technetium(IV) sulfide $TcS_2$ (j) ruthenium(III) iodide $RuI_3$

5. Write the formula for each of these Type II ionic substances that have the "older" name: (a) plumbic oxide $PbO_2$
(b) chromic chloride $CrCl_3$ (c) cupric sulfate $CuSO_4$ (d) stannous chloride $SnCl_2$
(e) mercurous nitrate $HgNO_3$ (f) stannic acetate $Sn(OAc)_4$ (g) plumbous dichromate $PbCr_2O_7$ (h) auric cyanide $Au(CN)_3$ (i) cobaltic sulfite $Co_2(SO_3)_3$ (j) aurous sulfate $Au_2SO_4$ (k) mercuric bromide $HgBr_2$

# Inorganic Chemistry: Introduction to Coordination Chemistry

6. Name the following binary Main-group compounds:
(a) CO  (b) $CO_2$  (c) $SO_3$  (d) $P_2O_5$  (e) $P_4O_{10}$  (f) NO
(g) $NH_3$  (h) $PCl_5$  (i) $PH_3$  (j) $SF_4$  (k) $XeF_4$  (l) $OF_2$

7. Write the formula for each of the following binary Main-group compounds: (a) xenon tetrafluoride $XeF_2$ (b) sulfur dioxide $SO_2$ (c) phosphorous pentachloride $PCl_5$ (d) diboron trioxide $B_2O_3$ (e) dinitrogen tetroxide $N_2O_4$ (f) silicon tetrabromide $SiBr_4$ (g) boron trifluoride $BF_3$ (h) nitric oxide $NO$ (i) nitrous oxide $N_2O$ (j) ammonia $NH_3$ (k) phosphine $PH_3$ (l) silicon dioxide $SiO_2$ (m) dichlorine oxide $Cl_2O$ (n) disulfur dichloride $S_2Cl_2$ (o) phosphorous trichloride $PCl_3$

8. What geometries are associated with coordination numbers 2, 4 and 6?
2: linear or bent
4: square planar; tetrahedral
6: octahedral; hexagonal planar; trigonal prismatic.

9. If one mole of the metal complex $[Co(NH_3)_6]Cl_3$ was dissolved in water, and then excess $AgNO_3$ (silver nitrate) was added to the solution, how many moles of AgCl would precipitate out of solution?   (a) zero  (b) 1  (c) 2  (d) 3 ✓

10. Conductivity measurements confirm that 1 mole of $CrBr_3 \cdot 4NH_3$ dissociates into 2 moles of ions in aqueous solution. For this octahedral metal complex write: $[Cr(NH_3)_4Br_2]Br$
    (a) the modern formula of the coordination complex and
    (b) an equation for the dissociation.
    $[Cr(NH_3)_4Br_2]Br \longrightarrow [Cr(NH_3)_4Br_2]^+ + Br^-$

11. For the octahedral complex $[Fe(H_2O)_6]^{3+}$, draw it and identify:
    (a) the coordination number of iron   6.
    (b) the coordination geometry of iron   octahedral
    (c) the oxidation state of the iron   $Fe^{3+}$

12. For the octahedral complex $[Mn(NH_3)_6]^{3+}$, draw it and identify:
    (a) the coordination number of manganese   6
    (b) the coordination geometry of manganese   octahedral
    (c) the oxidation state of manganese   $Mn^{3+}$

13. For the octahedral complex $[Co(NH_3)_5Cl]^{2+}$, draw it and identify:
    (a) the coordination number of cobalt   6

## Inorganic Chemistry: Introduction to Coordination Chemistry

(b) the coordination geometry of cobalt  octhedral
(c) the oxidation state of cobalt  $Co^{3+}$

14. For the octahedral complex $[Co(NH_3)_4Cl_2]^+$, draw it and identify:
    (a) the coordination number of cobalt  6
    (b) the coordination geometry of cobalt  octahedral
    (c) the oxidation state of cobalt  $Co^{3+}$

15. Combinations of cobalt(III), ammonia ($NH_3$), nitrite ($NO_2^-$) anions, and potassium ($K^+$) cations result in the formation of a series of seven octahedral coordination compounds.

(a) Write the modern formulas for the members of this series. (*Hint*: Not all seven compounds contain all four of the components.)

(b) How many ionic nitrites would there be in each compound?

(c) How many isomers would each compound have, assuming that each has an octahedral coordination sphere?

16. Combinations of iron(II), $H_2O$, $Cl^-$, and $NH_4^+$ can result in the formation of a series of seven octahedral coordination compounds, one of which is $[Fe(H_2O)_6]Cl_2$.

(a) Write the modern formulas for the other members of the series. (*Hint*: Not all seven compounds contain all four of the components.)

(b) How many chlorides could be precipitated from each compound by reaction with aqueous silver nitrate?

(c) How many isomers would each compound have, assuming that each has an octahedral coordination sphere?

17. Combinations of platinum(II) cations, ammonia molecules, thiocyanate anions, and ammonium cations result in the formation of a series of five square planar coordination compounds.

Write the modern formulas for the members of this series. (*Hint*: Not all five compounds contain each of the components.)

18. Combinations of palladium(II) cations, triphenylphosphine molecules, chloride anions, and ammonium cations result in the formation of a series of five square planar coordination compounds.

(a) Write the modern formulas for the members of this series. (*Hint*: Not all five compounds contain each of the components.)

(b) Which compound or compounds would have the greatest conductivity in solution? Briefly explain your answer.

(c) Write the formula for the neutral member of this series as Werner would have written it. Assuming that palladium(II) has a square planar coordination sphere, how many isomers would this compound have?

19. If dissolved in water, how many ions would be formed from dissolution of octahedral $[Co(NH_3)_5Cl_3]$ ?

   What are the ions?

   $[Co(NH_3)_5Cl_3]$  →

20. If dissolved in water, how many ions would be formed from dissolution of octahedral $[K_4FeCl_6]$ ?

   What are the ions?

   $[K_4FeCl_6]$  →

21. How many moles of Br- ion would be produced by dissolution of 1.5 moles of $[Co(NH_3)_5Br]Br_2$ in water?

22. If 1.96 grams of $[Co(NH_3)_6]Cl_3$ was reacted with an excess of silver nitrate, how much silver chloride (AgCl) would precipitate out of solution?

# Inorganic Chemistry: Introduction to Coordination Chemistry

23. If 2.54 grams of $Ba_3[FeCl_6]_2$ reacted with sodium sulfate $(Na_2SO_4)$ how much barium sulfate $(BaSO_4)$ would precipitate out of solution?

24. What would be the concentration (in molarity) of bromide ion in solution formed from 5.0 grams of $[Co(NH_3)_4Br_2]Br$ in 100ml of solution?

25. The octahedral rhodium (III) metal complex with the overall formula $[Rh(NH_3)_5Cl_3]$ was dissolved in water. What were the ions formed in aqueous solution?

$\quad\quad [Rh(NH_3)_5Cl_3] \quad \rightarrow$

26. The metal complex octahedral $(NH_4)_3[FeCl_6]$ was dissolved in water. What were the ions formed in aqueous solution?

$\quad\quad (NH_4)_3[FeCl_6] \quad \rightarrow$

27. Coordination compounds of formula $MA_4$ might be <u>square planar</u> or <u>tetrahedral</u>. How many isomers would you predict to exist for compounds of formula $MA_2B_2$ for these two geometries? $Pt(NH_3)_2Cl_2$ has two known isomers, and $[CoBr_2I_2]^-$ has only one.
    Speculate on the possible structures of these complexes by drawing them.

28. Using only a periodic table to locate the element, write the electron configuration for

(a) Os  (b) Co  (c) Ni  (d) Ru  (e) Cu  (f) Re

29. What is the highest possible oxidation state for each of the following transition metals?

(a) Zr  (b) Ta  (c) Mn  (d) Nb  (e) Tc  (f) Y

# Inorganic Chemistry: Introduction to Coordination Chemistry

30. Identify two transition metal ions with each of the following electron configurations:
    (a) [Ar] $3d^6$
    (b) [Ar] $3d^5$
    (c) [Ar] $3d^{10}$
    (d) [Ar] $3d^8$

31. Give the electron configuration for each of the ions listed.
    (a) $Pt^{2+}$
    (b) $Ni^{2+}$
    (c) $Co^{3+}$
    (d) $Ir^{3+}$
    (e) $Ti^{4+}$
    (f) $Zr^{4-}$
    (g) $Cu^{2+}$
    (h) $Ag^+$
    (i) $Cr^{3+}$
    (j) $Mo^{3+}$
    (k) $Mn^{3+}$
    (l) $Tc^{3+}$

32. What is the coordination number of ruthenium in $K_4[Ru(CN)_6]$?

33. What is the oxidation number of iridium in $[IrCl(NH_3)_5]Cl_2$?

34. In the d-block elements, the third-row metallic radii are about the same as the second-row radii because of ...
    (a) greater shielding of f-electrons.
    (b) greater shielding of d-electrons.
    (c) the lanthanide contraction.
    (c) the increased number of d-electrons.

35. Draw the structures for the isomers that are predicted for the three different geometries (hexagonal planar, octahedral, and trigonal prismatic) that are listed in Table 2. Advanced. (see Chapter 3. for bidentate ligands)

# Inorganic Chemistry: Introduction to Coordination Chemistry

36. Jörgensen synthesized [CoCl$_2$(en)$_2$]Cl that came in two forms named for their colors: violeo and praseo. Werner cited the existence of these two (and only two) isomers as proof of an octahedral coordination sphere.

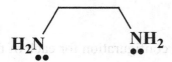

(a) Recalling that en stands for the bidentate ethylenediamine molecule, draw structural formulas for each isomer. (*Hint*: The ethylenediamine molecule can span only adjacent positions in the octahedron.)

(b) Suppose the coordination sphere for coordination number 6 had been trigonal prismatic instead of Werner's octahedral. How many isomers would this compound have had under that assumption? Draw structural formulas for each. (*Hint*: The ethylenediamine molecule can span only adjacent positions on the triangular and rectangular sides of the trigonal prism.)

(c) Suppose the coordination sphere for coordination number 6 had been hexagonal planar instead of Werner's octahedral. How many isomers would this compound have had under that assumption? Draw structural formulas for each. (*Hint*: The ethylenediamine molecule can span only adjacent positions on the hexagon.)

Bonus Questions

37. What is the formula weight of K$_4$[Ru(CN)$_6$] ?

38. What is the formula weight of [IrCl(NH$_3$)$_5$]Cl$_2$ ?

39. What is the formula weight of (NH$_4$)$_3$[FeCl$_6$] ?

40. If 25 mmol of CoCl$_3$ x 5 H$_2$O is reacted with excess aqueous ammonia to form [Co(NH$_3$)$_6$]Cl$_3$, how many grams of it would form if the yield is 33% ?

# Chapter 2. Lewis Electron Dot Structures and Valence Shell Electron Pair Repulsion Theory

## I. Lewis Electron Dot Structures

*Lewis Electron Dot Structures* were designed by *G.N. Lewis (1875-1946)* to show the comparison of the bonding power of an atom with the number of available (valence shell) electrons.

For example, carbon normally forms four bonds, and it has 4 available, or valence shell, electrons. Thus, it is drawn with 4 available electrons, whereas nitrogen has five available electrons but usually forms only three bonds. Most elements acquire an *octet* set of eight valence electrons (four pairs) in forming chemical bonds. The first and second row elements can be drawn with their representative Lewis Electron Dot Structures:

$$\overset{\cdot}{H} \quad \overset{\cdot\cdot}{He}$$

$$\overset{\cdot}{Li} \quad \overset{\cdot}{Be}\cdot$$

$$\overset{\cdot}{\underset{\cdot}{B}}\cdot \quad \cdot\overset{\cdot}{C}\cdot \quad \cdot\overset{\cdot\cdot}{\underset{\cdot}{N}}\cdot \quad \cdot\overset{\cdot\cdot}{\underset{\cdot\cdot}{O}}: \quad \cdot\overset{\cdot\cdot}{\underset{\cdot\cdot}{F}}: \quad :\overset{\cdot\cdot}{\underset{\cdot\cdot}{Ne}}:$$

The reaction of two chlorine atoms to form molecular chlorine gas and also the reaction to show the production of the ionic compound sodium chloride is shown using Lewis electron dot structures:

$$:\overset{\cdot\cdot}{\underset{\cdot\cdot}{Cl}}\cdot \quad + \quad \cdot\overset{\cdot\cdot}{\underset{\cdot\cdot}{Cl}}: \quad \longrightarrow \quad :\overset{\cdot\cdot}{\underset{\cdot\cdot}{Cl}}—\overset{\cdot\cdot}{\underset{\cdot\cdot}{Cl}}:$$

$$\overset{\cdot}{Na} \quad + \quad \cdot\overset{\cdot\cdot}{\underset{\cdot\cdot}{Cl}}: \quad \longrightarrow \quad Na^+ \quad :\overset{\cdot\cdot}{\underset{\cdot\cdot}{Cl}}:^-$$

In the first reaction shown above both Cl atoms acquire an "octet" of valence electrons and are thus "satisfied". In the

## Inorganic Chemistry: Introduction to Coordination Chemistry

second reaction, sodium loses its only valence electron, and Cl gains an electron so that both acquire a noble gas configuration (thus effectively acquiring an octet).

Lewis structures are used to show the molecular structures for simple molecules, both organic and inorganic, and are primarily used to describe compounds in organic chemistry and their reactions. Since ionic compounds aren't described as usefully with Lewis electron dot structures, then ionic compounds usually aren't depicted in this manner.

***VSEPR theory*** also utilizes Lewis electron dot structures to describe the simple molecular geometry in small organic and inorganic molecules.

**The rules used to draw Lewis electron dot structures** are given below:
(1) Sum up the total number of valence electrons for all the atoms in the molecule. For a polyatomic anion, add the number of negative charges to the total, and for a polyatomic cation subtract the number of positive charges from the total.
(2) Draw the skeletal structure for the molecule by connecting bonded pairs of atoms by a bond dash. Each ***bonding pair of electrons*** is denoted by a dash.
(3) Place electrons around the outer atoms so that each atom has an octet. *(The octet rule)* Hydrogen can only have a duet of electrons, and that must be its bonding pair.
(4) The remaining electrons (subtract the number assigned so far from the original total from step 1) are assigned in pairs to the central atom as ***lone pairs of electrons***.
(5) A multiple bond is likely if the central atom has fewer that an octet after step 4. A pair of electrons from an outer (peripheral) atom can be used to form a ***multiple bond***.

# Inorganic Chemistry: Introduction to Coordination Chemistry

*Additional Rules of thumb*:
(1) Central atoms are often written first in chemical formulas.
(2) Choose the atom with the lowest ionization energy for the central atom.
(3) Arrange atoms symmetrically around the central atom. ($SO_2$ is O-S-O, not S-O-O)

## Resonance Structures

More than one valid Lewis structure may be drawn for some molecules that have multiple bonds, and these are called *resonance structures*. An example is the nitrate ion $(NO_3)^-$. The following scheme shows the three possible resonance structures that can be drawn for the nitrate ion.

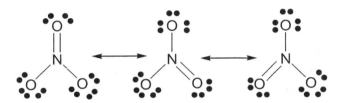

In the case of multiple bonded compounds such as nitrate, Lewis structures actually aren't a valid description of the molecule since there are no separate single and double bonds in the nitrate ion. **The x-ray crystal structure of the nitrate ion shows that all the N-O bond lengths are exactly the same length.** The pair of electrons that makes up the double bond between one of the N-O bonds is actually *delocalized* throughout the whole molecule. Thus, the bond order between the N and O atoms is actually one and 1/3, which points out an inherent failure in the use of Lewis structures. True chemical covalent bonds are spread over many atoms in molecular orbitals, and those will be discussed in a later chapter.

## Formal Charge

*Formal Charge* =
Valence e⁻s – (Lone Pair electrons + ½ of the number of electrons in bonds)

It may be hard to tell which atom is the central atom for certain linear molecules such as $CO_2$, and $N_2O$. Use of formal charge helps determine which structure is more likely. Structures with the lowest formal charges are likely to have the lowest energy, and therefore be the most stable. Also, the molecule is most stable when any negative charge resides on the most electronegative atom.

For example, the correct structure of $CO_2$ can be determined by calculating the formal charge for two likely structures. The formal charges are listed above the atoms:

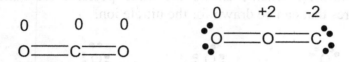

As can be seen, the structure with the carbon atom in the center has the lowest formal charges, and therefore the more stable structure. The formal charges for two possible structures for $N_2O$ are shown below:

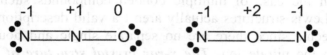

Of the structures shown above for $N_2O$, which do you expect to be the correct structure? (That's right, the 1$^{st}$ one!)

## Hypervalent compounds (Expanded Valence Shells)

Some molecules have central atoms that violate the octet rule and have more than four electron pairs around the central atom. For example:

$P_4 + 6\ Cl_2 \rightarrow 4\ PCl_3$    $PCl_3 + Cl_2 \rightarrow [PCl_4]^+ [PCl_6]^-$

and  $[PCl_4]^+ [PCl_6]^- \rightarrow 2\ PCl_5$

The complex ion [PCl$_6$]- has a phosphorous atom that violates the octet rule because it has six chlorides around it, each with a pair of bonding electrons for a total of twelve electrons around the central atom, and PCl$_5$ has ten.

The compound PCl$_5$, and its fluorine analogue PF$_5$ exhibit a trigonal bipyramid geometry which we will discuss shortly.

Nitrogen, which is a group V element like phosphorous, cannot form the analogous compounds NCl$_6$ or NF$_6$. Why?

Even though N and P are in the same group, it must be noticed that N is a 2$^{nd}$ row element, whereas P is a 3$^{rd}$ row element.

The simplest explanation is that phosphorous, since it is a 3$^{rd}$ row element, has available d orbitals that nitrogen does not; therefore using **valence bond theory**, nitrogen can only form sp$^3$ hybrid orbitals whereas P can form dsp$^3$ (five bonds) and d$^2$sp$^3$ (six bonds) hybrid orbitals and expand its octet.

Valence bond theory was put forth by Linus Pauling in his book "The Nature of the Chemical Bond" which was dedicated to G. N. Lewis. The next section briefly discusses the important points of valence bond theory with respect to geometry and molecular shape.

**Valence-Bond Theory**

Valence bond theory is based on the Lewis electron dot concept that covalent bonding results from sharing of electron pairs between atoms. The valence bond approach is most useful to organic chemists who use the theory very successfully.

The result of using the valence orbitals to form a mixed orbital (usually the s and the three p orbitals) is a *hybrid orbital*. The number of hybrid orbitals formed by this approach equals the number of atomic orbitals that are used to form the hybrid orbitals. Not only s and p orbitals can be used to form hybrid orbitals, but Pauling postulated that d orbitals could be used as well. The formation of hybrid orbitals using valence-bond theory can be used successfully to explain and account for molecular shapes.

**Table 2.1 Hybridization of orbitals and molecular shape.**

| Hybridization | # of hybrid orbitals | Molecule geometry |
|---|---|---|
| sp | 2 | Linear |
| $sp^2$ | 3 | Trigonal planar |
| $sp^3$ | 4 | Tetrahedral |
| $dsp^3$ | 5 | Trigonal bipyramidal |
| $d^2sp^3$ | 6 | Octahedral |

In Chapter 5 we will discuss the shortcomings of valence bond theory in dealing with transition metal complexes, and why it isn't used to explain the bonding and physical properties of these transition metal complexes.

**Limitations of VSEPR Theory**

The majority of molecules that have a molecular geometry that is trigonal bipyramid or octahedral are hypervalent compounds. An exception to this is the transition metal coordination compounds that we have discussed in Chapter 1., and these transition metal coordination compounds cannot be adequately described by VSEPR theory.

An example of this breakdown of Lewis electron dot structures to explain the geometry of metal complexes can be found in the simplest Werner complexes. For example, the cobalt(III) compound $Co(NH_3)_6^{3+}$ would not have octahedral molecular geometry around the central metal atom. Cobalt (III) has six valence electrons itself, plus twelve more electrons from the six ammonia ligands, giving a total of 18 electrons around the central atom. A total of 18 electrons around the central atom would have nine pairs of electrons around the central atom and would not be octahedral.

Therefore, we will limit the use of Lewis electron dot structures and VSEPR theory to explain the molecular and electronic geometry of only simple molecules and ligands.

**Electron Deficient Compounds**

Some molecules have central atoms that have less than an octet of electrons. These compounds are called *electron deficient* compounds. Examples of electron deficient compounds are often found with central atoms of the elements boron, beryllium, and aluminum. An example is $BF_3$, boron trifluoride, which is considered to be a *Lewis acid*. (See Chapter 3. for more discussion on *Lewis Acids* and *Lewis Bases*) Other examples are $BeH_2$, and trimethyl aluminum.

Examples:

Two valid Lewis structures for the very reactive molecule $H_2BF$ are shown below. One structure shows that boron does not have an octet of electrons, whereas the second structure does.

Calculate the formal charge on each atom in the structures. Which structure is more reasonable in your opinion? Why?

## II. The VSEPR Theory:
## Electronic Geometry and Molecular Shapes.

Valence Shell Electron Pair Repulsion Theory (VSEPR) is a practical, though imperfect means of predicting the basic geometry of many covalent molecules and molecular ions. Many molecules have one atom that is bonded to two or more other atoms. These atoms are called *central atoms*. Atoms that are bonded to only one other atom are *outer* or *peripheral atoms*. This is important in two respects, one is that many simple inorganic molecules are best described in this way, and two, the central atom theme compares well to the metal atom geometries found in transition metal coordination compounds.

A central atom will be surrounded by a number **of regions of high electron density (RHED)**. There are two types of RHED, shared and unshared. An unshared RHED is simply a lone pair of electrons. A shared RHED can be two electrons (a single bond), four electrons (a double bond), six electrons (a triple bond), or a non-integral number of electrons (a resonance bond). For the purposes of VSEPR, all shared RHED are the same. *Most* of the time the RHED is simply an electron pair-- hence the name of the theory is "Valence Shell Electron Pair Repulsion Theory", and not "Valence Shell Region of High Electron Density Repulsion Theory".

VSEPR is based on the following principles:

**Principle 1**: The RHED around a central atom repel one another. Thus, the RHED form angles in which they are as far from one another as possible.

**Principle 2**: Unshared RHED ("lone pairs") take up a little more room than shared RHED.

From these two principles, one can qualitatively, and sometimes fairly quantitatively, predict bond angles in an impressive variety of molecules and molecular ions.

# Inorganic Chemistry: Introduction to Coordination Chemistry

## III. Lewis Electron Dot Structures

The first step in using VSEPR is to determine the Lewis electron dot structure of the molecule. This will answer two fundamental questions essential to applying VSEPR--how many RHED surround the central atom and how many of these are unshared RHED?

## IV. Applying VSEPR

One can speak of two kinds of geometry that exist simultaneously in a molecule or molecular ion, the electronic geometry and the molecular geometry. *Electronic geometry* is the geometry of all RHED around the central atom, whether shared or unshared. There are only 5 common electronic geometries. These correspond to 2, 3, 4, 5, and 6 RHED around the central atom.

*Molecular geometry* is the geometry of the atomic centers in the molecule. Thus, the molecular geometry, while it is influenced by the electronic geometry, is determined by considering how many shared RHED and how many unshared RHED are around the central atom.

### A. Two Regions of High Electron Density: Linear geometry

Imagine an atom with two RHED around it. Applying Principle 1, these repel each other and try to get as far apart as possible. It is not hard to imagine that this is achieved by forming a bond angle of 180°, which gives a linear geometry.

The most common example of linear geometry is $CO_2$. In the electron dot structure, each oxygen is double bonded to the carbon atom, O=C=O. (We'll ignore unshared pairs on the outer atoms--they aren't important to the geometry.) The two RHED around the carbon (double bonds in this case) repel each other and this repulsion causes the molecule to become *linear*. Because both RHED are shared, the molecular geometry is also linear.

Another example is hydrogen cyanide, HCN. The electron dot structure, H:C:::N, has a single bond between the H and the C, and a triple bond between the C and the N (again ignoring unshared pairs on outer atoms, in this case on the nitrogen). Once again, the RHED (a single and a triple bond) repel, and as in $CO_2$, both the electronic and molecular geometry is "linear".

$$\ddot{\ddot{O}}=C=\ddot{\ddot{O}} \qquad H-C\equiv N:$$

## B. Three Regions of High Electron Density

1. No unshared pairs, **$AB_3$**
(A = central atom, B = peripheral atom)

Imagine two RHED around a central atom. Now imagine bringing one more RHED in. Because of the mutual repulsion, the two original RHED will bend away from the incoming RHED, forming a triangle of RHED. The triangle lies in a plane, and so the electronic geometry with three RHED is known as *trigonal planar*.

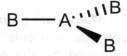

**Trigonal Planar**

For example, consider the electron deficient compound $BF_3$. The Lewis dot structure tells us we have three RHED around the B, three single bonds. All the RHED are shared so the molecular geometry is also trigonal planar. All three F and the B lie in the same plane, and the F-B-F bond angles are all $120°$.

The same goes for $NO_3^-$, a molecule that exhibits resonance. In this case there are again three RHED, only this time each resonance structure has one double and two single bonds. What's important, however, is that three RHED are present, and so the geometry is still trigonal planar.

2. One unshared pair, **AB₂U**

(A = central atom, B = peripheral atom, U = unshared pair)

Now, suppose one of the RHED is unshared (a lone pair). This occurs in $SO_2$, where the Lewis structure indicates a lone pair on the sulfur, and a resonance single/double bond between the sulfur and each oxygen atom. Once again, the electronic geometry is trigonal planar, but the molecular geometry, the geometry of the atoms, is *bent*. The angle of bending is *about* 120°, but because the unshared RHED takes a little more room, the bond angle bends to slightly less than 120°, (about 119.5°). Note the difference between $SO_2$ and $CO_2$. $CO_2$ is linear, but $SO_2$ is bent because of the presence of the lone pair on the sulfur atom.

**Bent**

## C. Four Regions of High Electron Density

This is a very common case, because if all four are single bonds, then the octet rule is satisfied for the central atom.

Go back to our trigonal planar situation, with, for example, three electron pairs around a central atom. Now bring in a fourth electron pair perpendicularly.

The three originals will be repelled, and will bend out of the plane. The resulting geometry is that of a "tetrahedron". So, the electronic geometry when four RHED are present is ***tetrahedral***. The bond angles in the tetrahedron are all ~109.5°.

## 1. No unshared electron pairs: **AB₄**

If there are no unshared RHED, the molecular geometry is also tetrahedral. This is the case for SiCl₄. The four RHED around the silicon (electron pairs in this case) form a tetrahedron. All Cl-Si-Cl bond angles are exactly 109.5°, and all four chlorine's are in exactly the same positions relative to the silicon and the other three chlorine atoms.

**Tetrahedral**

## 2. One unshared electron pair: **AB₃U**

This is another common case. Consider NH₃. The four RHED around nitrogen (three single bonds and one unshared pair) form a tetrahedral electronic geometry, but the three shared RHED (with the hydrogen's on the end) will form a triangle, like in BF₃. But unlike BF₃, the central atom is **not** in the same plane as the three hydrogen atoms. This will look like a three-sided pyramid. This molecular geometry is called *trigonal pyramidal*. The unshared RHED will take up a bit more space, so the bond angle is somewhat less than 109.5°. (In NH₃, the bond angle is 107.3°)

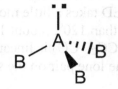

**Trigonal Pyramid**

## 3. Two unshared RHED: **AB₂U₂**

Another common case is H₂O. The four RHED in water are two single bonds and two unshared pairs on the oxygen. While the electronic geometry is still tetrahedral, the molecular geometry is **bent**. In the case of SO₂, which was bent, the bond angle was just a little less than 120°, the trigonal planar bond angle. The unshared RHED in an AB₂U₂ case compress the H-O-H bond angle to something less than 109.5°. In the case of water, this angle is 104.5°.

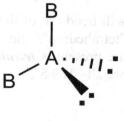

**Bent**

## D. Five Regions of High Electron Density

With five RHED there are at least 10 electrons around the central atom, so now these compounds exhibit extended valence shells, which can occur in some atoms in the third row and later rows. These are hypervalent compounds with an expanded octet. To envision what will happen with five RHED, go back to the tetrahedron, and think of four electron pairs around a central atom. Now bring a fifth electron pair in perpendicular to the triangle formed by three of the original RHED. The fifth pair will force three of the originals back into the same plane. Now we have a set of three RHED forming a trigonal plane, and two more RHED forming a line perpendicular to the trigonal plane. The shape is like two trigonal pyramids placed base to base. This is called a *trigonal bipyramid*, the electronic geometry with five RHED.

**Trigonal Bipyramid**

Notice that we have two distinct kinds of RHED: the three in the trigonal plane, and the two perpendicular to the plane. The three in the trigonal plane are called "equatorial", and the two perpendiculars are called "*axial*". The equatorial positions have a different geometric relationship to the rest of the positions than do the axial positions.

1. No unshared pairs: **AB$_5$**

If there are no unshared RHED, the molecular geometry is the same as the electronic geometry, trigonal bipyramid. Probably the most famous example is PCl$_5$.
Three of the Cl's are in the trigonal plane, forming 120°

**Trigonal Bipyramid**

angles with each other. Two of the Cl's are axial, forming a bond angle of 180° with each other. The equatorial Cl form 90° angles with the axial Cl.

2. One unshared pair: **AB₄U**

Now in determining the molecular geometry a new problem is encountered. There are two different positions for the unshared pair: equatorial and axial.

By Principle 2 we know that unshared pairs occupy more space than shared bonding pairs of electrons, but which position---axial, or equatorial--- provides the optimum space? It turns out that there is *more space in an equatorial position than an axial position*.

**See-Saw**

Therefore, the unshared pair will be in one of the equatorial positions.

An example of a molecule like this is $SF_4$. While the formula is something like $CF_4$, the lone pair on the sulfur gives it a different geometry. Two of the F are equatorial; the other two F are axial.

This unusual molecular geometry has been called **_disphenoidal_** or *see-saw*. The two equatorial F form an angle of a bit less than 120° with each other, and the two axial F form a bond angle of a bit less than 180°.

3. Two unshared pairs: **AB₃U₂**

Once again, we choose to put the unshared pair in an equatorial position. An example of this is $ClF_3$. One of the F is in the remaining equatorial position; the other two are axial.

The molecular geometry is ***T-shaped***.

**T-Shaped**

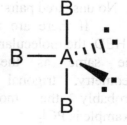

## 4. Three unshared pairs: $AB_2U_3$

The third unshared pair goes into the final equatorial position.
An example is $KrF_2$, one of the very few krypton compounds.
Both F are in the axial position, so the molecular geometry is strictly linear
Therefore, $KrF_2$ and $BeCl_2$ have the same molecular geometry, but for entirely different reasons.

**Linear**

## E. Six Regions of High Electron Density

**Octahedral**

The final case we'll consider here is six RHED around the central atom. Once again, in our imagination we'll start with the trigonal bipyramid. Imagine bringing yet another electron pair, this time bisecting two of the equatorial RHED. These two original equatorial RHED will be forced to move away, and in the end we'll have an *octahedral* electronic geometry.

The octahedral looks like two four-sided pyramids (square pyramids) back-to-back, and could be (but is not) called "square bipyramidal".

### 1. No lone pairs: $AB_6$

This molecular geometry is octahedral. An example is $SF_6$. All six F are entirely equivalent to each other, each forming a 90° angle with four other F, and an 180° angle with the fifth one.

**Octahedral**

## 2. One lone pair: $AB_5U$

All positions in an octahedron are equivalent (unlike the trigonal bipyramid), and so we don't have to think about where the lone pair will go. It will simply assume one of the positions, and the remaining five atoms will lie in a molecular geometry, which is called *square pyramidal*.

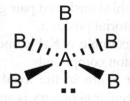

**Square Pyramid**

An example is the rare inorganic molecule $IF_5$.

## 3. Two lone pairs: $AB_4U_2$

Now there are two possible choices for the second lone pair: it can either be at 90° to the other lone pair, or at 180° to the other lone pair. Applying Principle 2, one assumes that the two lone pairs will be as far apart as they can be, and that will be at 180°. An example is the noble gas compound $XeF_4$. The two lone pairs on Xe are on opposite sides of a plane of Xe and the four F atoms.

**Square Planar**

The four F atoms form a molecular geometry called *square planar*. The F atoms define the square, and the Xe lies directly in the center of the plane.

## F. Summary

In the basic study of valence shell electron pair repulsion theory for common molecules using Lewis electron dot structures we therefore have five electronic geometries, and thirteen molecular geometries.

| System | Total RHED | Unshared RHED | Shared RHED | Electronic Geometry | Molecular Geometry |
|---|---|---|---|---|---|
| $AB_2$ | 2 | 0 | 2 | Linear | Linear |
| $AB_3$ | 3 | 0 | 3 | Trigonal Planar | Trigonal Planar |
| $AB_2U$ | 3 | 1 | 2 | Trigonal Planar | Bent |
| $AB_4$ | 4 | 0 | 4 | Tetrahedral | Tetrahedral |
| $AB_3U$ | 4 | 1 | 3 | Tetrahedral | Trigonal pyramidal |
| $AB_2U_2$ | 4 | 2 | 2 | Tetrahedral | Bent |
| $AB_5$ | 5 | 0 | 5 | Trigonal Bipyramidal | Trigonal Bipyramidal |
| $AB_4U$ | 5 | 1 | 4 | Trigonal Bipyramidal | See-saw |
| $AB_3U_2$ | 5 | 2 | 3 | Trigonal Bipyramidal | T-shaped |
| $AB_2U_3$ | 5 | 3 | 2 | Trigonal Bipyramidal | Linear |
| $AB_6$ | 6 | 0 | 6 | Octahedral | Octahedral |
| $AB_5U$ | 6 | 1 | 5 | Octahedral | Square pyramidal |
| $AB_4U_2$ | 6 | 2 | 4 | Octahedral | Square Planar |

## Electronegativity, Polarity, and Polar Molecules

In the 1930's and in his book *"The Nature of the Chemical Bond"* author and Nobel Prize winner **Linus Pauling** analyzed the energies of chemical bonds and he developed an *Electronegativity* scale. Electronegativity is defined as the ability of an atom to attract electrons to itself in a chemical bond. Atom electronegativities are useful in deciding if a chemical bond is polar, and which atom has a partial negative charge and which has a partial positive charge. The following scheme shows electronegativity values.

**Pauling Electronegativity Values**

| 2.2 H | | | | | | | |
|---|---|---|---|---|---|---|---|
| 1.0 Li | 1.6 Be | 2.0 B | 2.5 C | 3.0 N | 3.5 O | 4.0 F | |
| 0.9 Na | 1.3 Mg | 1.6 Al | 1.9 Si | 2.2 P | 2.6 S | 3.1 Cl | |
| 0.8 K | 1.0 Ca | 1.8 Ga | 2.0 Ge | 2.2 As | 2.6 Se | 3.0 Br | |
| 0.8 Rb | 0.9 Sr | 1.8 In | 1.9 Sn | 2.0 Sb | 2.1 Te | 2.7 I | |
| 0.8 Cs | 0.9 Ba | | | | | | |

One of the most confusing aspects of inorganic chemistry is the blurring of the line between ionic and covalent compounds. By looking at the periodic table and the associated electronegativity values of the elements we can predict if a compound has primarily covalent or ionic bonds.

If the electronegativity values of two atoms bonded together are exactly the same, we say it is a ***purely covalent bond***. The only way that this occurs is to have two atoms of the same element bonded together, as in for example, $Br_2$, $O_2$, or $N_2$.

If the electronegativity value difference between two bonding atoms is near 1.0-2.0, then we say it is an *ionic bond*. Examples would be LiF, NaCl, HF, etc. These are true salts or strong acids and bases.

If the electronegativity difference values are low, between 0 and 0.5, then we say the bond is a *polar covalent bond*, but not purely covalent. One of the atoms has a larger electronegativity value than the other, and electron density is found to be greater on the atom with the highest electronegativity. This causes bonds to be polar.

Sometimes polar covalent bonds are arranged in a symmetrical fashion around the central atom, and we find that the molecule <u>does not</u> have a dipole moment. If the polar covalent bonds are unsymmetrically arranged around the central atom, we find that the molecule does have a *dipole moment*.

## Transition Metal Electronegativity Values

The transition metals have a very narrow range of electronegativity values that range from 1.3 for scandium to 1.9 for some of the heavier metals like Os, Ir, and Au. The electronegativity scale is therefore not as useful in describing the properties of transition metals as it is in describing the properties of ligands.

### Pauling Electronegativity Values

| 3 | 4 | 5 | 6 | 7 | 8 | 9 | 10 | 11 | 12 |
|---|---|---|---|---|---|---|---|---|---|
| 21 Sc 1.3 | 22 Ti 1.4 | 23 V 1.5 | 24 Cr 1.6 | 25 Mn 1.6 | 26 Fe 1.7 | 27 Co 1.7 | 28 Ni 1.8 | 29 Cu 1.8 | 30 Zn 1.6 |
| 39 Y 1.3 | 40 Zr 1.3 | 41 Nb 1.5 | 42 Mo 1.6 | 43 Tc 1.7 | 44 Ru 1.8 | 45 Rh 1.8 | 46 Pd 1.8 | 47 Ag 1.6 | 48 Cd 1.6 |
| 71 La 1.3 | 72 Hf 1.3 | 73 Ta 1.4 | 74 W 1.5 | 75 Re 1.7 | 76 Os 1.9 | 77 Ir 1.9 | 78 Pt 1.8 | 79 Au 1.9 | 80 Hg 1.7 |

The low electronegativity values for the early transition metals means that they do not form as covalent an interaction with carbon in low oxidation states as do the later metals.

## Dipole Moments

Certain molecules, in the gas phase, can be aligned in an electric field. Examples of these molecules are hydrogen fluoride and water. We should ask ourselves why this occurs...why are we able to align these molecules in an electric field?

**HF Molecules Aligned in an Electric Field**

$$\left[ \begin{array}{c} \text{H—F} \\ \text{H—F} \\ \text{H—F} \\ \text{H—F} \\ \text{H—F} \\ \text{H—F} \end{array} \right]$$

−    +

Hydrogen fluoride is a polar molecule, and this means that it has a dipole moment. The electronegativity of fluorine (4.0) is much greater than that of hydrogen (2.1), and thus the covalent bond between H and F is polar covalent since the difference in their electronegativity values is 1.9.

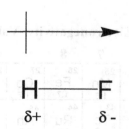

H—F
δ+    δ-

The arrow points to the negative end of the polar bond, and the + end of the arrow designates the positive end. The arrow indicates the polar nature of the bond, and also of the entire molecule since it is linear.

HF is not an ionic compound, therefore the negative end is considered to have a *"partial negative charge"*, which is designated with a ***delta minus*** (δ -) symbol, and the positive end likewise is considered to have a "*partial positive charge*" and is designated with a ***delta plus*** (δ+) symbol.

This is a case where the molecule contains a polar bond, and the molecule itself is polar and has a dipole moment. It is not always the case that a molecule contains polar bonds and is itself polar! Let's look at such a case in the next example.

To illustrate this point we will compare HF to another linear molecule, carbon dioxide or $CO_2$. Carbon dioxide has double bonds between carbon and oxygen as can be seen in the structural representation shown next.

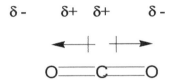

Each C=O bond is polar. However the polar bond of one C=O bond cancels out the effect of the other C=O bond with a net effect of <u>no polarity</u> for the entire molecule itself. (You can imagine these arrows as force vectors that cancel out)

The symmetrical nature of the molecule is the key to understanding if a molecule is polar or not. To have symmetry in a molecule means that parts of it can be interchanged with others without altering the identity or the orientation of the molecule. Symmetrical molecules, where any given one side of the molecule is the same as the other, are not polar, whereas unsymmetrical molecules are always polar.

The best way to systematically study polar molecules is to work up through the different molecular geometries. Since we have looked at linear molecules we should therefore examine trigonal planar molecules next, and then tetrahedral molecules.

The trigonal planar molecule boron trifluoride is not considered to be a polar molecule even though it contains three polar bonds. This is illustrated in the representation of the molecule $BF_3$ drawn below.

# Inorganic Chemistry: Introduction to Coordination Chemistry

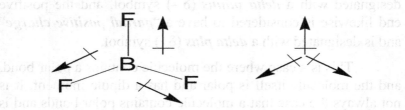

Again, this molecule is symmetrical and one-half of the molecule is like the other half.

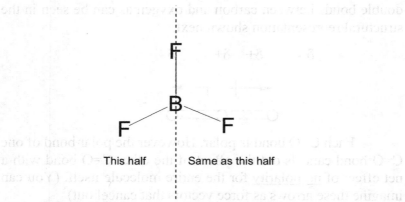

It is instructive to examine a planar molecule ($BF_2Cl$) that is very similar to $BF_3$.

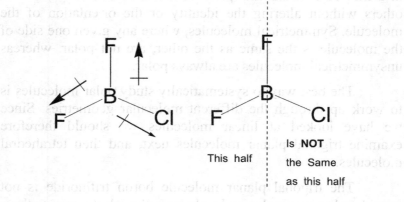

Notice that the arrow showing the polar bond between B and Cl is not as large in the drawing shown above as that of the polar bond between B and F.

This signifies that the B-Cl bond is not as polar as the B-F bond, and this can be calculated using the electronegativity values for each element.

The values are: B=2.0 , Cl=3.0 , and F= 4.0. Obviously the B-F bond is much more polar.

Next, the molecule is definitely unsymmetrical. As shown above, one-half of the molecule is not the same as the other half. With all these things considered, the $BF_2Cl$ molecule shown above is a polar molecule and should exhibit a dipole moment, even if it is very small.

The scope of this book is limited, and is meant as an introduction to coordination chemistry on the sophomore college level. Thus, we use symmetry elements on a basic level, without introducing symmetry and group theory the way that it should really be presented (which would take an entire book to cover adequately) so the symmetry discussions in this section are meant to be basic and elementary.

Let us examine the next molecular geometry, the tetrahedral electronic geometry, which is more complicated.

Three molecules come easily to mind, methane ($CH_4$), methylene chloride ($CH_2Cl_2$) and carbon tetrachloride ($CCl_4$).

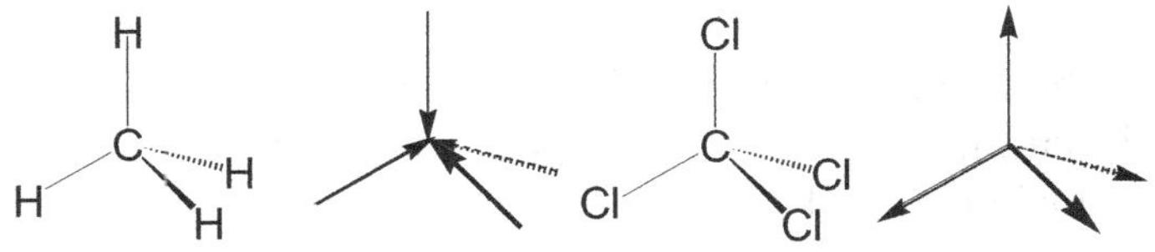

Methane and carbon tetrachloride are non-polar molecules because they have no dipole as can be seen in the previous scheme.

However, methylene chloride and chloroform do exhibit a dipole moment, and they are classified as ***polar organic molecules***.

The arrows in the following scheme depict polar covalent bonds, and also the overall dipole.

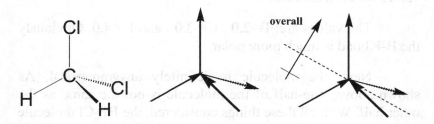

Dichloromethane has one side that is different from the other, creating an overall dipole moment.

The same is true for chloroform, but the dipole is slightly different.

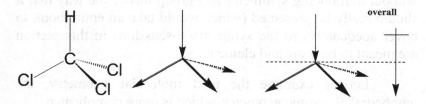

What about chloromethane $CClH_3$? Does that molecule have a dipole moment?

Yes, it does.

The trigonal bipyramid molecular geometry is actually a combination of two geometries in one. That is, there is a linear portion (axial), and also a trigonal planar portion (equatorial). Molecules that contain different substituents on the axial and equatorial positions are rare, but the mental exercise to see if a proposed molecule would have a dipole moment can be instructive.

The prototype molecule is PF$_5$, shown below.

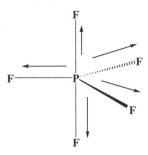

The molecule has no dipole moment as the vectors for all the polar bonds cancel out. If a Cl atom, in either an axial or an equatorial position, replaces one of the F atoms on PF$_5$ then the resulting molecule would be polar, with a weak dipole.

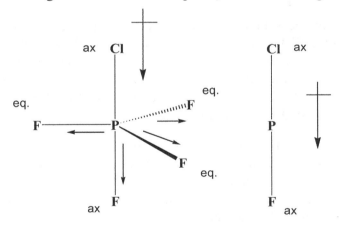

In the scheme above, the axial (ax) and the equatorial (eq) substituents on the molecule are labeled. Since the equatorial substituents are all F, with polar P-F bonds, they cancel each other out.

On the other hand, the axial substituents are different, making one side of the molecule different from the other. The P-F bond is more polar than the P-Cl bond, so the dipole lies in the direction indicated.

If the Cl were placed on an equatorial position instead, the following situation would occur as depicted in the next scheme.

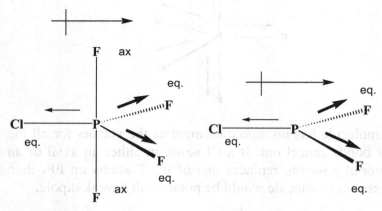

In the scheme above, since the axial substituents are the same, all F, with polar P-F bonds, then they cancel each other out. On the other hand, the equatorial substituents are different, making one side of the molecule different from the other. The P-F bond is more polar than the P-Cl bond, so the dipole lies in the direction indicated.

The disphenoidal geometry is, by its very unsymmetrical nature, prone to exhibit a dipole moment in all compounds that adopt that structure. For example the molecule $SF_4$ has a dipole moment, because we can draw a plane through the molecule where one-half is different from the other.

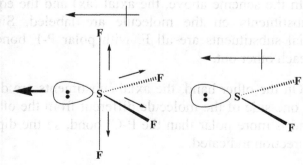

**Disphenoidal**

Molecules with a T-Shaped molecular geometry are, of course, polar molecules. $ClF_3$ is such an example, and the following scheme depicts the molecule. As in the previous example, lone pairs of electrons are considered to be regions of high electron density (RHED), so they are considered strongly when proposing dipole moments.

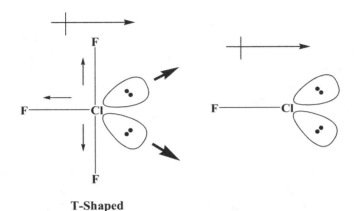

**T-Shaped**

The octahedral geometry is an interesting case since it could potentially involve many of the transition metal compounds that we may observe. Again, it is the breaking of perfect symmetry about the molecule that is the key to observation of a dipole moment. The molecule $SF_6$ is the prototype molecule of octahedral molecular geometry for a non-transition metal compound, and it is not a polar molecule, as can be seen in the scheme below.

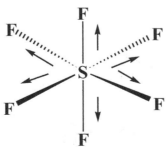

If one examines a metal complex that does not have this perfect symmetry around the central atom, for example $[Co(NH_3)_5Cl]^{2+}$, then the molecule does exhibit a dipole moment.

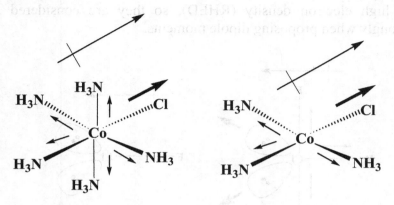

This is also the case with cis-$[Co(NH_3)_4Cl_2]^+$, where it does have a dipole moment, but this is not the case for trans-$[Co(NH_3)_4Cl_2]^+$, which does not have a dipole moment.

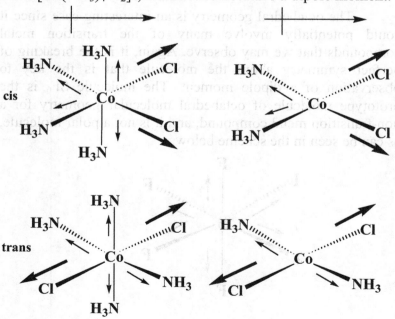

If a compound is examined with an octahedral electronic geometry such as $IF_5$, which is a rare example of a non-transition metal compound with a square pyramid molecular geometry, one sees that it has a dipole moment.

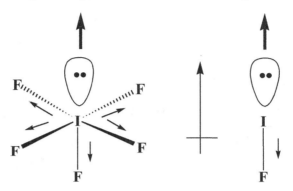

The square planar molecular geometry is basically a special case of the octahedral molecular geometry. In the square planar case of non-transition metal compounds such as the rare $XeF_4$ molecule, there is no dipole moment.

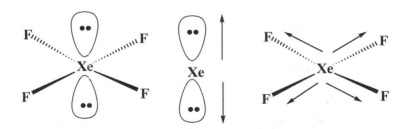

The electron pairs cancel each other out (axial positions) and the Xe-F bonds cancel each other out (equatorial positions) leaving a molecule with no discernible dipole moment.

In the case of transition metal complexes that do not have lone pairs associated with the central atom in a VSEPR sense, the situation is different, but there are similarities. For example, $[PtF_4]^{2-}$ has no dipole moment, but $[Pt(NH_3)_3Cl]^-$ does.

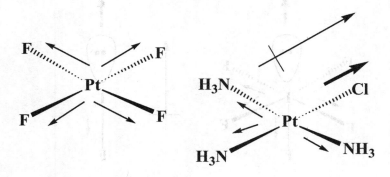

The molecule cis-platin has a dipole moment. (Compare to cis-$[Co(NH_3)_4Cl_2]^+$) The isomer of cis-platin, the trans-platin molecule, has no dipole moment. The cis-and trans-platin molecules are shown in the scheme below.

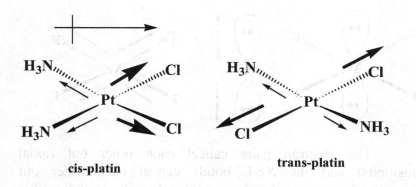

cis-platin                    trans-platin

# Rare Geometries

There are higher coordinate geometries than octahedral (six coordinate) but they are rare, with few examples. Even coordination number six has a rare representative, the trigonal prismatic geometry.

The best example of this rare geometry is $Re[S_2C_2(C_6H_5)_2]_3$, a rhenium dithiolate complex, where the dithiolate ligand is a ***bidentate ligand*** (see Chapter 3), which forces the Re into this geometry.

Seven, eight, and nine coordinate metal complexes are known. Usually, to get this many ligands around a central atom, the ligands need to be small (like $F^-$) and the metal has to be fairly large (like a 2$^{nd}$ or 3$^{rd}$ row transition metal, or a rare earth or actinide metal).

| Coord # | Geometry | Example |
|---|---|---|
| 7 | Pentagonal bipyramid | $ZrF_7^{2-}$, $V(CN)_7^{4-}$ |
| 7 | Monocapped octahedron | $NbF_7^{2-}$, $TaF_7^{2-}$, |
| 7 | Monocapped trigonal prism | $NbOF_6^{3-}$ |
| 8 | Cubic | $Na_3[PaF_8]$, $Na_3[UF_8]$ |
| 8 | Square antiprism | $ReF_8^{2-}$, $W(CN)_8^{4-}$ $Zr(acac)_4$ |
| 8 | Dodecahedron | $Mo(CN)_8^{4-}$ |
| 8 | Hexagonal bipyramid | $[CdBr_2(18\text{-crown-}6)]$ |
| 9 | Tricapped trigonal prism | $ReH_9^{2-}$ |

# Inorganic Chemistry: Introduction to Coordination Chemistry

## High Coordinate Geometries

**7-coordinate**  **7-coordinate**  **7-coordinate**
Pentagonal Bipyramid   Monocapped   Octahedron

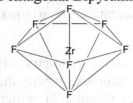  

Monocapped Trigonal Prism

**8-coordinate**  **8-coordinate**  **8-coordinate**
Cubic   Square Antiprism

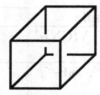  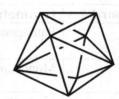

Dodecahedron

**8-coordinate**  **9-coordinate**

Hexagonal Bipyramid   Tricapped Trigonal Prism

 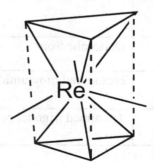

# Inorganic Chemistry: Introduction to Coordination Chemistry

## Terms and Definitions  Chapter 2

*Lewis Electron Dot Structures* –

*G.N. Lewis* --

*VSEPR Theory* –

*The Octet Rule* –

*Bonding Pairs Of Electrons* –

*Lone Pairs Of Electrons* –

*Multiple Bond* –

*Resonance Structures* –

*Delocalized* –

*Formal Charge* –

*Hypervalent Compounds* –

*Electron Deficient* –

*Lewis Acid* –

*Lewis Base* –

*Central Atoms* –

*Peripheral Atoms* –

*Regions Of High Electron Density (RHED)* –

*Electronic Geometry* –

*Molecular Geometry* –

*Linear* –

*Trigonal Planar* –

*Tetrahedral* –

## Inorganic Chemistry: Introduction to Coordination Chemistry

*Trigonal Pyramidal* –

*Bent* –

*Trigonal Bipyramid* –

*Disphenoidal (Seesaw)* –

*T-Shaped* –

*Axial And Equatorial Sites* –

*Octahedral* –

*Square Pyramid* --

*Square Planar* –

*Linus Pauling* --

*The Nature Of The Chemical Bond* --

*Electronegativity* –

*Valence Bond Theory* ---

*Purely Covalent Bond* –

*Ionic Bond* --

*Polar Covalent Bond* --

*Dipole Moment* –

*Partial Negative Charge* --

*Partial Positive Charge* --

*Polar Organic Molecules*

# Inorganic Chemistry: Introduction to Coordination Chemistry

## Chapter 2 Problems and Exercises

1. Which of the following species can have resonance structures?
   - (A) $O_3$
   - (B) $BF_4^-$
   - (C) $CO_3^{2-}$
   - (D) $NO_2^-$
   - (E) $CH_3NH_2$

2. How many resonance forms can be written for the nitrate ion?

3. How many lone pairs of electrons does the nitrogen atom possess in the Lewis structure of HCN?

4. For the Lewis structure below, the formal charges on N, C, and S, respectively, are...
$$[::C=S=N::]^-$$

5. For the Lewis structure below, the formal charges on C, S, and N, respectively, are....
$$[::N=C=S::]^-$$

6. All of the following have Lewis structures that obey the octet rule except ...
(A) NO   (B) $NO_2^+$   (C) $N_2O_4$   (D) $N_2O$   (E) $N_3^-$

7. The number of lone pairs around the central iodine atom in $I_3^-$ is _____.

8. Which of the following are Lewis acids?
$$CN^-, Al^{3+}, Cl^-, PCl_3, BF_3$$

9. What is the molecular shape of $AlCl_4^-$?

10. What is the molecular shape of $IF_2^-$?

# Inorganic Chemistry: Introduction to Coordination Chemistry

11. What is the molecular shape of $SO_3$?

12. The F–S–F bond angle in $SF_2$ is ...

13. What is the molecular shape of $CO_3^{2-}$?

14. The Cl–P–Cl bond angle(s) in $PCl_5$ is/are ....

15. Which one of the following molecules is polar?
Draw them all.
    (A) COS  (B) $S_2$  (C) $CO_2$  (D) $CS_2$  (E) $O_2$

16. Which one of the following molecules is polar?
Draw them all.
    (A) $CH_2F_2$  (B) $XeF_4$  (C) $CH_4$  (D) $BF_3$  (E) $CCl_4$

17. All of the following molecules are polar except...
Draw them all.
    (A) $SF_2$  (B) $XeF_2$  (C) $OF_2$  (D) $SF_4$  (E) NOCl

18. Which of the following ions have noble gas electron configurations? List the valence shell electronic configurations for all of them.
    (a) $Fe^{2+}$, $Fe^{3+}$, $Sc^{3+}$, $Co^{3+}$
    (b) $Tl^+$, $Te^{2-}$, $Cr^{3+}$
    (c) $Pu^{4+}$, $Ce^{4+}$, $Tl^{3+}$
    (d) $Ba^{2+}$, $Pt^{2+}$, $Mn^{2+}$

19. Write Lewis electron dot structures that obey the octet rule for each of the following:
    (a) CN-    (b) CO    (c) $PH_3$    (d) $PF_3$    (e) SCN-


20. Although both $Br_3^-$ and $I_3^-$ ions are known, the $F_3^-$ ion has not been observed.
    Why not?

# Inorganic Chemistry: Introduction to Coordination Chemistry

| System | Total RHED | Unshared RHED | Shared RHED | Electronic Geometry | Molecular Geometry |
|---|---|---|---|---|---|
| $CS_2$ | | | | | |
| $BF_3$ | | | | | |
| $CO_3^{2-}$ | | | | | |
| $O_3$ | | | | | |
| $SO_4^{2-}$ | | | | | |
| $PCl_3$ | | | | | |
| $ClO_3^-$ | | | | | |
| $H_2S$ | | | | | |
| $PF_5$ | | | | | |
| $PF_4^-$ | | | | | |
| $ICl_3$ | | | | | |
| $XeF_2$ | | | | | |
| $PF_6^-$ | | | | | |
| $BrF_5$ | | | | | |
| $XeF_4$ | | | | | |

# Inorganic Chemistry: Introduction to Coordination Chemistry

**Practice Quiz**  *DRAW STRUCTURES!!!* That will help you!

____1. Which of these molecules do contain polar bonds?
(a) $H_2O$  (b) $H_2S$  (c) $CO$  (d) $CO_2$  (e) all contain polar bonds

____2. Which of these molecules do not contain polar bonds?
(a) $H_2$  (b) $N_2$  (c) Ar  (d) $I_2$  (e) none contain polar bonds

____3. Which of these molecules DOES NOT have a dipole moment?
(a) $H_2$  (b) $BF_3$  (c) Ar  (d) $I_2$  (e) none have a dipole moment

____4. Which of these molecules DOES have a dipole moment?
(a) $CH_4$  (b) $XeF_4$  (c) $O_2$  (d) $CO_2$  (e) $NH_3$

5.-9. Match the compound with its correct ***electronic geometry*** (one each)
(a) $CO_2$  (b) $H_2O$  (c) $PCl_5$  (d) $XeF_4$  (e) $NO_2^-$
____5. Tetrahedral       ____6. Trigonal Bipyramidal
____7. Octahedral
____8. Trigonal Planar
____9. Linear

____10. Triple bonds count as only a single pair of electrons when deciding the correct electronic geometry around the central atom of a molecule.   (a) true  (b) false

11.-15. Match the compound with its correct ***molecular geometry*** (one each)
(a) $CH_4$  (b) $H_2O$  (c) $PCl_5$  (d) $XeF_4$  (e) $SF_4$

____11. Tetrahedral       ____12. Trigonal Bipyramidal
____13. Square Planar     ____14. Bent
____15. See-Saw

# Inorganic Chemistry: Introduction to Coordination Chemistry

_____16. The electronic geometry of the central atom in $PCl_3$ is:
(a) pyramidal  (b) trigonal planar  (c) tetrahedral
(d) octahedral  (e) trigonal bipyramidal

_____17. The **molecular geometry** of $PCl_3$ is:
(a) pyramidal  (b) trigonal planar  (c) tetrahedral
(d) octahedral  (e) trigonal bipyramidal

_____18. Four of the following statements about the ammonia molecule, $NH_3$, are correct.
One is not correct. Which is it?
(a) The ammonia molecule has tetrahedral molecular geometry.  (b) Since nitrogen is more electronegative than hydrogen, the bond dipoles are directed toward the nitrogen atoms.
(c) The bond dipoles re-enforce the effect of the unshared pair of electrons on the nitrogen atom.  (d) The bond angles in the ammonia molecule are less than $109°$.

_____19. The **molecular shape** of $SO_3$ is:
(a) pyramidal  (b) trigonal planar  (c) tetrahedral
(d) octahedral  (e) trigonal bipyramidal

_____20. **The molecular shape** of $AlCl_4^-$ is:
(a) pyramidal  (b) trigonal planar  (c) tetrahedral
(d) octahedral  (e) trigonal bipyramidal

_____21. The Lewis dot formula for $Br_2$ shows _____.
(a) a single covalent bond.    (b) a double covalent bond.
(c) a triple covalent bond.    (d) a single ionic bond.
(e) a total of 8 x 2 = 16 electrons.

_____22. The number of unshared pairs of electrons in the outer shell of sulfur in $H_2S$ is _____.
(a) one   (b) two   (c) three   (d) four   (e) zero

# Inorganic Chemistry: Introduction to Coordination Chemistry

___23. The number of unshared pairs of electrons in the outer shell of arsenic in $AsF_3$ is ___.
(a) one  (b) two  (c) three  (d) four  (e) zero

___24. Which molecule exhibits resonance?
(a) $BeI_2$  (b) $O_3$  (c) $H_2S$  (d) $PF_6$  (e) $CO_2$

___25. How many lone pairs of electrons are there on the Xe atom in the $XeF_4$ molecule?
(a) one  (b) two  (c) three  (d) four  (e) zero

___26. The elements of Group VIIA may react with each other to form covalent compounds. Which of the following single covalent bonds in such compounds is the most polar bond?
(a) F-F  (b) F-Cl  (c) F-Br  (d) F-I  (e) Cl-I

___27. Which one of the compounds below has the bonds that are the most polar?
(a) $H_2S$  (b) $PH_3$  (c) $AsCl_3$  (d) $SiH_4$  (e) $SbCl_3$

___28. Which one of the following molecules does not have a dipole moment?
(a) BrCl  (b) FCl  (c) FBr  (d) $O_2$  (e) ClBr

# Chapter 3. Ligands and Nomenclature

The transition metal studies of Werner were followed later by the concepts of G. N. Lewis and N. V. Sidgwick, who recognized that formation of a chemical bond required the sharing of an electron pair. This led to the idea that a molecule or ion with an electron pair can donate these electrons (definition of a *Lewis base*) to a metal ion, which is considered to be an electron acceptor (definition of a *Lewis acid*). The definition of a *ligand* is any molecule or ion that has <u>at least one electron pair that can be donated to a metal</u>.

$$H^+ \;+\; [:\!\ddot{O}\!-\!H]^- \;\longrightarrow\; H_2O$$

**Lewis Acid**　　**Lewis Base**

$$BF_3 \;+\; :NH_3 \;\longrightarrow\; BF_3\!-\!NH_3$$

**Lewis Acid**　　**Lewis Base**　　**Coordinate Covalent Bond**

Ligands are *Lewis bases* since they donate a pair of electrons to the metal and are attracted to them, thus they are also called *nucleophiles*. Molecules such as $BF_3$ with incomplete valence electron shells, or metal ions in a metal complex, are *Lewis acids* since they are electron acceptors, and they are also *electrophiles* since they seek out ligands.

When one atom donates both electrons to form a covalent bond, it is termed a *coordinate covalent bond*. Coordinate covalent bonds are formed between ligands and metals when metal complexes are formed.

$$Ni^{2+} \;+\; 4\,[:\!\ddot{\underset{..}{Cl}}\!:]^- \;\longrightarrow\; [NiCl_4]^{2-}$$

**Lewis Acid**　　**Lewis Base**

$$Ni^{2+} \;+\; 4\;:NH_3 \;\longrightarrow\; [Ni(NH_3)_4]^{2+}$$

**Lewis Acid**　　**Lewis Base**

## Hard and Soft Acid-Base Concepts

The use of hard and soft acid-base concepts is more of a rule of thumb that is useful in describing the general tendencies of certain metals *(Lewis Acids)* to bond preferentially to certain ligands (*Lewis Bases*) than it is a real working theory.

In general there is a tendency for ligands to bind to metals to form the most stable complex, so that a ligand that binds weakly to a metal may be replaced by a ligand that binds more strongly. (L* is a more strongly binding ligand than L in this example)

$$M\text{-}L + L^* \leftrightarrow M\text{-}L^* + L$$

A ranking of the bonding of ligands with a certain metal species can be determined in order to construct a ranking of ligand Lewis base strengths. This has been done, of course, and it turns out that the ranking of ligand Lewis base strengths depends on the metal.

In general, ligands such as $NH_3$, $OH^-$, $Cl^-$, $F^-$, $O^{2-}$, and $H_2O$ tend to form strong complexes with metals such as $Cr^{3+}$, $Fe^{3+}$, $Co^{3+}$, and $Ti^{4+}$. These ligands all have ligating atoms (O, N, etc.) that are all very electronegative and small. The metals are also very small, thus the ligands and metals are both considered ***nonpolarizable*** and ***hard***.

On the other hand, ligands such as $CN^-$, $PR_3$, $AsR_3$, $SR_2$, $SR^-$, and $CO$ tend to form strong complexes with metals such as $Cu^+$, $Hg^+$, $Cd^{2+}$, and $Au^+$. These ligands all have ligating atoms (S, P, etc.) that have a lower electronegativity than the hard ligands discussed above, and they are larger. The metals are also much larger, thus the ligands and metals are both considered ***polarizable*** and ***soft***.

The terms soft and hard are very imprecise but the concept is useful for guessing the stability of metal complexes with various ligands. The ambiguity is often due to the large number of borderline cases of ligands and metals.

**Ligands can be classified in several ways.**
**(A) Ligands may be classified by bonding mode**
**(B) Ligands may be classified by # of donated electrons**
**(C) Ligands may be classified as neutral, anions or cations**
**(D) Ligands may be classified by structural type**

(A) There are *two main classes of ligand bonding types* according to Cotton and Wilkinson (in "Advanced Inorganic Chemistry"), and for the general introductory study of transition metal coordination compounds this is surely the **most important way to classify ligands**. The reason that this is the most important way to classify ligands is not apparently obvious to the student of coordination chemistry at this point, but later, after the discussion of crystal field theory and molecular orbital theory, the importance will become apparent. The two bonding modes are discussed below, along with representative ligands.

**(1) Classical or simple σ-donor ligands** act as electron-pair donors to acceptor ions or molecules, and form complexes with all types of metal ions. These classical ligands are common neutral or anionic ligands such as chloride ions, water molecules, and ammonia molecules, but they may be more complicated ligands such as EDTA. These ligands are simple sigma (σ) electron pair donors.

**(2) Nonclassical ligands or π-bonding-π-acid -ligands** form coordination compounds only with transition metal atoms. There are three major ligand groups of this type:

*(a) **triply bonded molecules***: carbon monoxide (CO), cyanide $(CN)^-$, nitrosyl ion $(NO^+)$, isocyanides (CN-R)

*(b) **unsaturated organic molecules*** like alkenes and alkynes, benzene or the cyclopentadienyl anion,

*(c) **organophosphine ligands*** such as triphenylphosphine and bis(diphenylphosphino)ethane (includes arsines).

Non-classical ligands utilize a form of π-bonding to enhance their sigma donor bonding to a metal ion, thereby increasing the overall strength of the metal-ligand bond.

Unfortunately, as will be discussed later in the text, the borderline between sigma and pi bonding between ligands and metals is not as clear as one would like it to be.

$$\text{H}_2\text{N-CH}_2\text{-CH}_2\text{-NH}_2 \quad \longrightarrow \quad \underset{M}{\text{H}_2\text{N-CH}_2\text{-CH}_2\text{-NH}_2}$$

**(B) A third way of classifying ligands is by their electrical charge. Ligands are <u>neutral, anions or cations.</u>**

By convention, neutral ligands do not interfere with the oxidation state of the metal. On the other hand, ~~anionic~~ ligands oxidize the metal by one for every negative charge they possess, and cationic ligands ~~decrease~~ the oxidation state of the metal by one for every positive charge they possess. This is the ICC basis for electron counting.

**(C) Ligands may also be classified according to the number of electrons that they contribute to a central atom.**

This is a matter of *convention* on how the electrons are counted. Are the electrons counted as belonging to the ligand or the metal...or do they belong to both? The differences cause confusion.

There are two conventions for counting electrons in metal complexes. The most important convention, the one that is used most often and is the easiest to understand, is the <u>Ionic Charge Convention</u> (ICC method).

## *The Ionic Charge Convention (ICC Method)*

The basic premise of this metal-complex electron counting method is that we first remove all of the ligands from the metal with the proper number of electrons for each ligand to bring it to a ***closed valence shell state***, which is usually an octet of electrons around the central atom as we learned in Chapter 2.

For example, if we remove water from a metal complex, $H_2O$ has a completed octet and acts as a neutral molecule. When it was bonded to the metal center it did so through its lone pair and there is no need to change the oxidation state of the metal to balance charge. We therefore call water a neutral two-electron donor.

By contrast however, if we remove a chlorine group from a metal complex and complete its octet, then we formally have chloride ion ($Cl^-$). If we bond this chloride anion back to the metal, a lone pair forms our metal-chlorine bond and the chloride ion acts as a two-electron donor.

*Anionic ligands* are considered to be two electron donors by the ICC method -examples are $Cl^-$, $OH^-$, and $CH_3^-$. These two-electron donors have a negative charge since they have, in effect, oxidized the metal by one oxidation state.

On the other hand, anionic ligands that can form a covalent bond are considered to be *one-electron donors* by the Neutral Covalent Convention (NCC) method. To be counted as a one-electron donor means that one considers that the ligand is bonding covalently with the metal, not in an ionic fashion. Later, we will examine both of these methods (ICC method and the NCC method) in counting electrons in Chapter 5.

The ICC method is preferred by most chemists and **does not** distinguish anionic ligands as one-electron donors but as two-electron donors. We will use the ICC method in this text, except when we are comparing the two methods.

Any compound with an electron pair is a *two-electron donor.* Common examples are neutral molecules such as ammonia ($:NH_3$) or water ($H_2O$).

Groups that can form a single bond and at the same time donate a pair of electrons can be considered to be either four-electron donors by the ICC method, or more rarely, as three-electron donors by the NCC method. For example, the acetate ion can be either a two or four electron donor by the

ICC method, or ~~one~~ or a ~~three~~ electron donor using the NCC method.

|  | ICC Method | NCC Method |
|---|---|---|
| acetate ion — Metal | 2 electron donor | 1 electron donor |
| acetate ion — Metal | 4 electron donor | 3 electron donor |

A molecule with two electron pairs (for example, ethylene diamine, $H_2NCH_2CH_2NH_2$) can be regarded as **a *four-electron donor*,** and so on.

By convention, neutral ligands do not interfere with the oxidation state of the metal.

On the other hand, anionic ligands oxidize the metal by ~~one~~ for every negative charge they possess, and cationic ligands ~~decrease~~ the oxidation state of the metal by one for every positive charge they possess. Again, this is the ICC basis for electron counting.

**(D) A fourth way of classifying ligands is *structurally*,** which is by the number of individual bonding connections that the ligand makes to the central metal atom.

When only one atom becomes bonded the ligand is said to be ***monodentate*** or ***unidentate*,** such as the ligands in $Co(NH_3)_6^{3+}$, $NiCl_4^{2-}$, or $Ru(CN)_6^{3-}$. Thus, even though the ammonia ($NH_3$) ligand has four atoms, there is only one atom which bonds to the metal; the ligating atom is therefore nitrogen (N).

When a ligand uses two atoms to bond to a metal center, it is ***bidentate***. This is the case for the tertiary organophosphine ligands, dppe= bis(diphenylphosphino)methane, and dppe= bis(diphenylphosphino)ethane, shown below:

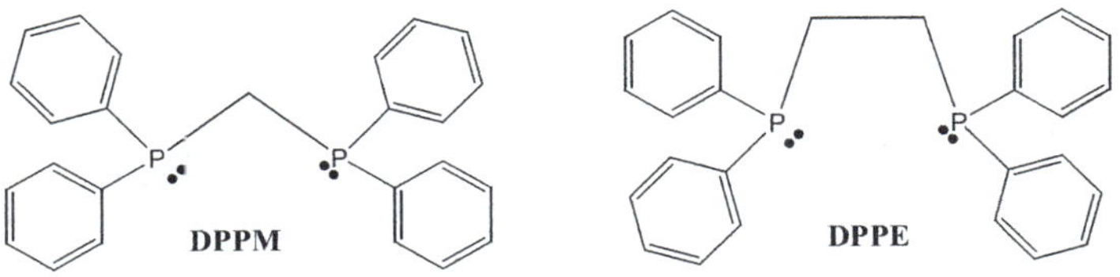

[Structures of DPPM and DPPE]

The organophosphine ligands, as depicted in the last pages of Table 3.1, are a very important class of ligands for transition metal complexes, the structures of which can be tailored to suit the needs of the chemist.

Bidentate ligands bonded to only one metal atom are termed ***chelate***, *(key-late--from the Greek--like a crab's claw)* as in the example using the ***bidentate chelating*** ethylene diamine ligand, or as shown below with the dppe ligand.

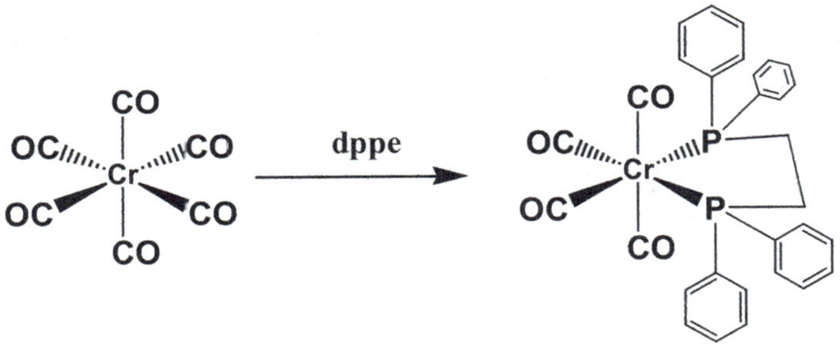

[Reaction scheme: Cr(CO)$_6$ + dppe → Cr(CO)$_4$(dppe)]

A ligand may also be ***tridentate, tetradentate,*** or ***hexadentate***, and so on, these are all ***multidentate ligands***.

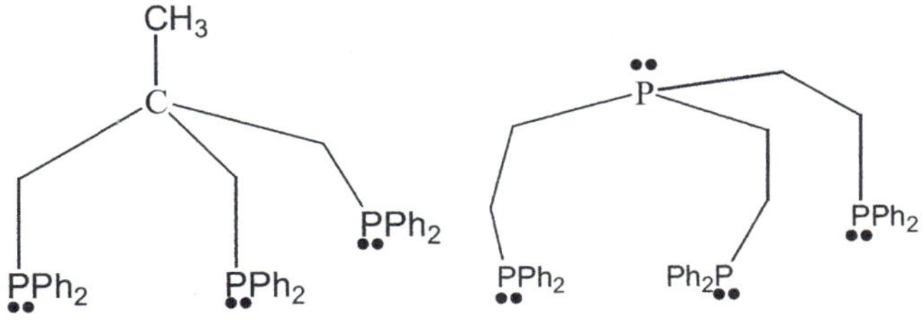

[Structures of tridentate and tetradentate phosphine ligands]

91

The multidentate phosphine ligands shown above can act as tridentate and tetradentate ligands respectively.

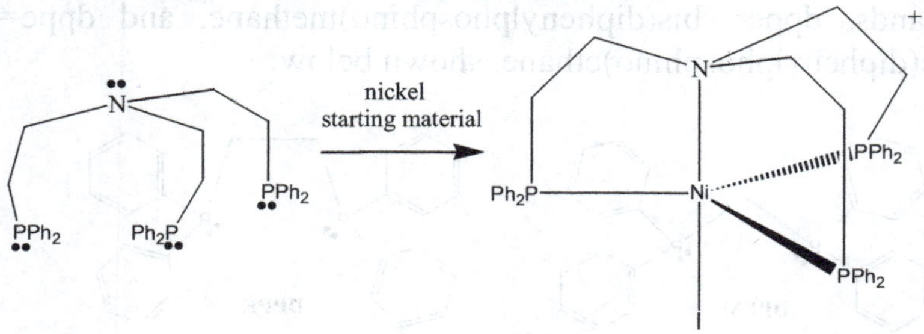

In this case, the ligand helps stabilize an odd geometry for the nickel metal complex, the trigonal bipyramid.

In the next example nickel is in a more common 4-coordinate geometry with another tetradentate ligand, which in this case is a planar ligand called a Schiff base.

The ligand coordinates around either Nickel(II) or Copper(II) to form a slightly tetrahedrally distorted square planar complex. The [$N,N'$-Bis(5-bromosalicylidene)-1,3-diaminopropane]nickel(II) complex is formed by coordination of the nickel (II) ion with the four-coordinate $N_2O_2$ donor set of the Schiff base imine-phenol ligands.

In this Ni compound, the Ni-O and Ni-N distances are 1.908 (3) and 1.959 (4) Å, respectively.

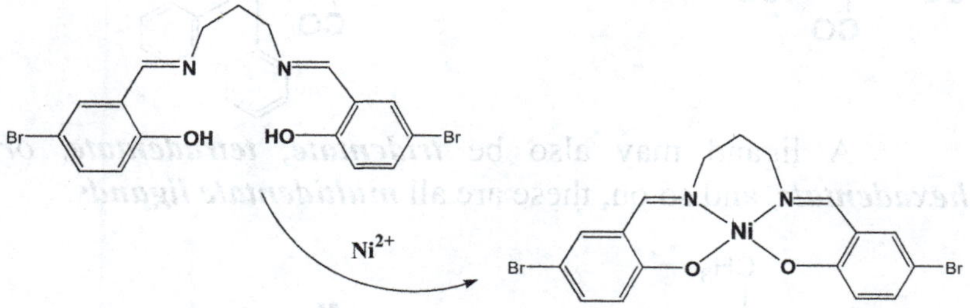

# Inorganic Chemistry: Introduction to Coordination Chemistry

The following schemes show how multidentate ligands bind with a transition metal through their ligating atoms (in bold).

### en is a bidentate ligand

### dien is a tridentate ligand

### trien is a tetradentate ligand

### EDTA can be a **hexadentate** ligand

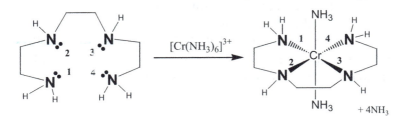

## The Chelate Effect

Chelating ligands often form very stable compounds with metal atoms because of *inherent thermodynamic stability*. This thermodynamic stability is called the *chelate effect*, and has enthalpic and entropic components.

The entropic component is by far the most important portion of the chelate effect and can be seen in the following scheme:

$[Ni(NH_3)_6]^{2+} + 3\text{ en} \longrightarrow [Ni(en)_3]^{2+} + 6\text{ NH}_3$

**4 molecules          produce          7 molecules**

The reaction produces more molecules than were originally reacted together, thus increasing the entropy ($\Delta S$) and contributing to a negative free energy ($\Delta G$). Remember the equation learned in general chemistry: $\Delta G = \Delta H - T\Delta S$, where $\Delta G$ is Gibbs free energy, $\Delta H$ is enthalpy, and $\Delta S$ is entropy. If $\Delta G$ is negative, then the process is spontaneous.

If one "arm" of the ligand is knocked away from the metal atom, in a collision in solution for example, then the other arm holds the chelate together long enough for the other arm to reattach itself to the metal. This is an enthalpic stability in that the total energy of the chelate metal-ligand bond is double that of a comparable monodentate ligand.

Many ligands act as *bridging groups*. In many cases single atoms or anions act as *unidentate bridging* ligands. This means that there is only *one* ligand atom that forms two or three bonds to different metal atoms. Larger, multi-atom ligands can sometimes act as *bidentate bridging* ligands, as seen in the following examples.

**Unidentate Bridging**            **Bidentate Bridging**

For monoatomic ligands, such as the chloride ion, and ligands containing only one donor atom, the ***unidentate*** form of bridging is the only possible mode.

Bridging ligands can be used to construct complexes and organometallic catalysts that exhibit interesting properties. In the following example, the rhodium complex has an open space between the Rh atoms that provides for interesting catalytic activity.

The accepted nomenclature notation for bridging ligands is the prefix descriptor η (eta) and for chelating ligands it is the prefix descriptor μ (mu). The prefix $η^n$ indicates that a ligand is using n # of its atoms to form bonds to metal atoms. For example, the EDTA ligand when binding in a hexadentate fashion as shown on page 93 can be described as $η^6$-EDTA. For molecules with an obvious denticity, such as dppe or en, the eta descriptor is left off, as the denticity and # of ligating atoms is understood. The prefix μ indicates that a ligand bridges only two metal atoms. If a ligand bridges three, or four metal atoms, the descriptors $μ^3$, or $μ^4$, and so on, are used.

Some unidentate ligands have two or more different donor sites so that the possibility of *linkage isomerism* arises. Some important ligands of this type, which are called **ambidentate ligands**, are:

M-$NO_2$  Nitro         M-ONO  Nitrito
M-SCN  Thiocyanato      or S-Thiocyanato
M-NCS  Isothiocyanato   or N-Thiocyanato

An example of two O- and N- bonded cobalt nitrito complexes are shown in the following scheme. Interestingly, the two forms are in equilibrium.

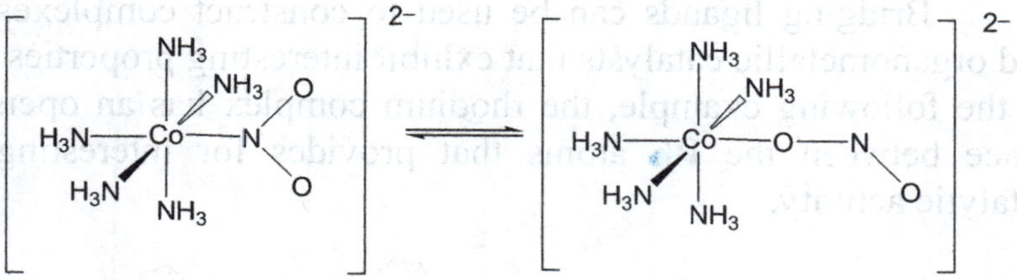

*Heterobimetallic* --------This term describes a complex in which there are two different metal centers, for example, ruthenium (Ru) and molybdenum (Cr).

*Homobimetallic*--------These complexes have two metal centers that are the same element. The two metals centers don't need to have identical ligands or coordination number, but are often found as symmetric dimers.

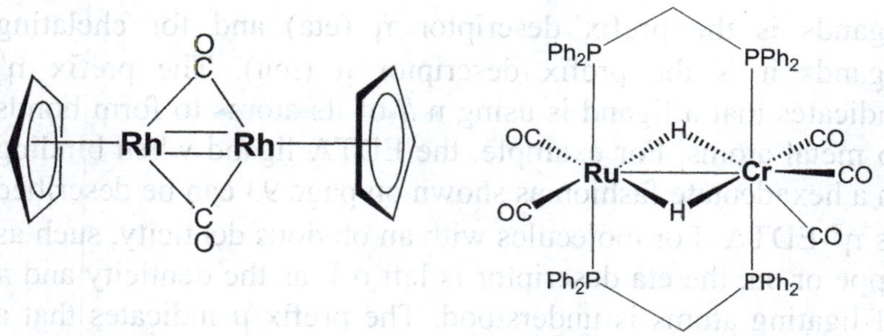

**Homobimetallic**                **Heterobimetallic**

Ligands may also be *macrocyclic* in nature. These have even more thermodynamic stability than typical chelating ligands because of the rigid nature of the ligand itself.

An example is for the *porphyrin* ring shown below, and an example of a porphyrin analog binding to an atom of magnesium to form chlorophyll A. Other examples in nature find such porphyrin rings structures bound to Fe, Co, and Zn to form important biological enzymes.

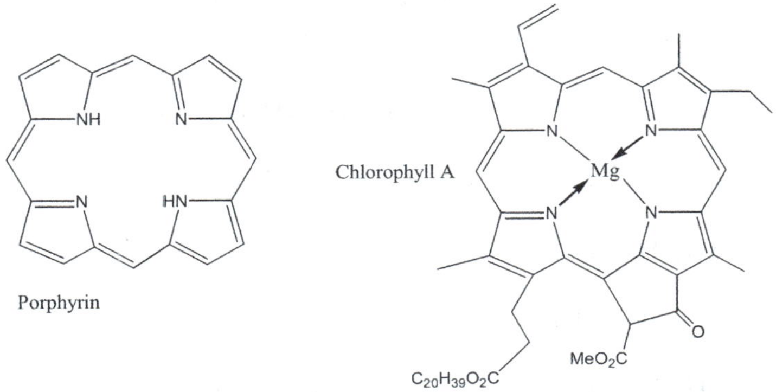

Chlorophyll A

Porphyrin

$C_{20}H_{39}O_2C$

$MeO_2C$

The **macrocyclic effect** is observed when a macrocyclic ligand replaces other ligands, including chelating ligands with the same denticity.

The macrocyclic effect is due to movement of the open arms in the open form compared to the lack of mobility for the closed macrocycle, which also gives more thermodynamic stability to the closed macrocycle.

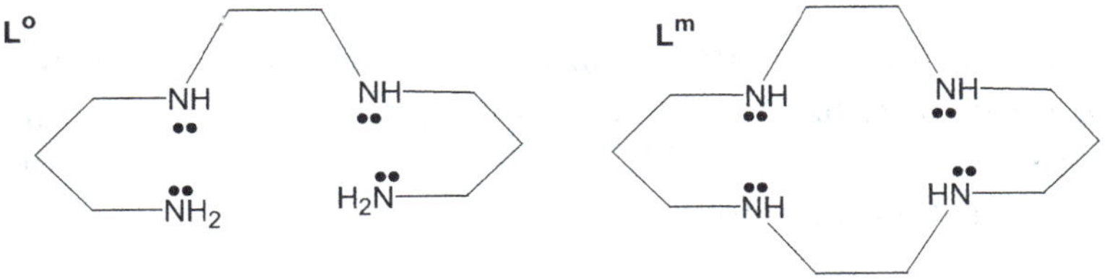

$L^o$

$L^m$

$Cu^{2+}_{(aq)} + L^o \rightarrow [CuL^o]^{2+}$ and then

$[CuL^o]^{2+} + L^m \rightarrow [CuL^m]^{2+} + L^o$

## Thermodynamic Stability and Formation Constants

The thermodynamic stability of a coordination compound is often expressed by the equilibrium constant for the reaction of the aqueous metal ion, such as $[Co(H_2O)_6]^{3+}$, with the reacting ligand. For example:

$$[Co(H_2O)_6]^{3+} + 6\,NH_3 \leftrightarrow [Co(NH_3)_6]^{3+} + 6\,H_2O$$

This process proceeds by a stepwise displacement of all the water molecules with new ammonia ligands until all six have been replaced.

($K_1$) $[Co(H_2O)_6]^{3+} + NH_3 \leftrightarrow [Co(H_2O)_5(NH_3)]^{3+} + H_2O$

($K_2$) $[Co(H_2O)_5(NH_3)]^{3+} + NH_3 \leftrightarrow [Co(H_2O)_4(NH_3)_2]^{3+} + H_2O$

($K_3$) $[Co(H_2O)_4(NH_3)_2]^{3+} + NH_3 \leftrightarrow [Co(H_2O)_3(NH_3)_3]^{3+} + H_2O$

($K_4$) $[Co(H_2O)_3(NH_3)_3]^{3+} + NH_3 \leftrightarrow [Co(H_2O)_2(NH_3)_4]^{3+} + H_2O$

($K_5$) $[Co(H_2O)_2(NH_3)_4]^{3+} + NH_3 \leftrightarrow [Co(H_2O)(NH_3)_5]^{3+} + H_2O$

($K_6$) $[Co(H_2O)(NH_3)_5]^{3+} + NH_3 \leftrightarrow [Co(NH_3)_6]^{3+} + H_2O$

The overall equilibrium expression for this reaction, generally referred to as $\beta_n$ (where n=6 for this reaction), is defined as:

$$\beta_6 = \frac{[Co(NH_3)_6]^{3+}}{[Co(H_2O)_6]^{3+}[NH_3]^6}$$

The $\beta_6$ is the *overall stability constant* or the *overall formation constant*. Water does not appear in the equilibrium expression. The values of K and β are related since β is the overall constant, therefore:

$\beta_6 = K_1 \times K_2 \times K_3 \times K_4 \times K_5 \times K_6$ and

$\log \beta_6 = \log K_1 + \log K_2 + \log K_3 + \log K_4 + \log K_5 + \log K_6$.

## Table 3.1 of Common Ligands

| Common Name | Abbreviation | |
|---|---|---|
| fluoro, | $F^-$ | |
| chloro, | $Cl^-$ | |
| bromo, | $Br^-$ | |
| iodo, | $I^-$ | |
| cyano, | $CN^-$ | |
| thiocyano, | $SCN^-$ | |
| isothiocyano, | $NCS^-$ | |
| hydroxo, | $OH^-$ | |
| aqua, | $H_2O$ | |
| carbonyl, | $CO$ | |
| thiocarbonyl, | $CS$ | |
| nitrosyl, | $NO^+$ | |
| nitro, | $NO_2^-$ | |
| nitrito, | $ONO^-$ | |
| oxalato, | $C_2O_4^{2-}$ or ox | |
| phosphine, | $PR_3$ | |
| pyridine, $C_5H_5N$ | pyr | |
| ammine, $NH_3$ | | |
| methylamine, $CH_3NH_2$ | | |
| ethylenediamine, $NH_2CH_2CH_2NH_2$, | | en |
| diethylenetriamine, $NH_2C_2H_4NHC_2H_4NH_2$ | | dien |
| triethylenetetramine, $NH_2C_2H_4NHC_2H_4NHC_2H_4NH_2$ | | trien |
| β, β', β"-triaminotriethylamine, $N(C_2H_4NH_2)_3$ | | tren |
| acetylacetonato, $[CH_3COCHCOCH_3]^-$ | | acac |
| 2,2'-bipyridine, | | bipy |
| 1,10-phenanthroline, $C_{12}H_8N_2$ | | phen |
| dialkyldithiocarbamate, $S_2CNR_2-$ | | dtc |
| 1,2-bis(diphenylphosphino)ethane, $PPh_2C_2H_4,PPh_2$ | | dppe |
| ethylenediaminetetraacetate, $(-OOCCH_2)_2,NCH_2CH_2N(CH_2COO-)_2$, | | EDTA |

# Structures and Abbreviations

**acac**, **en**, **ox**, **bipy**, **trien**, **phen**, **2, 3, 2-tet**, **dtc**, **EDTA**, **cyclam**

Note: The ligands <u>acac, ox, dtc and EDTA</u> are shown as their deprotonated anions here in this table, not in their neutral protonated forms.

# Inorganic Chemistry: Introduction to Coordination Chemistry

## Monodentate Organophosphine Ligands

**phosphine**

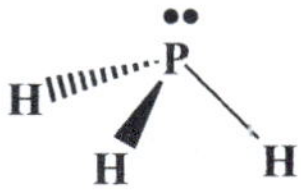

**trimethylphosphine**

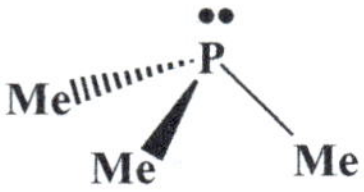

**triphenylphosphine**

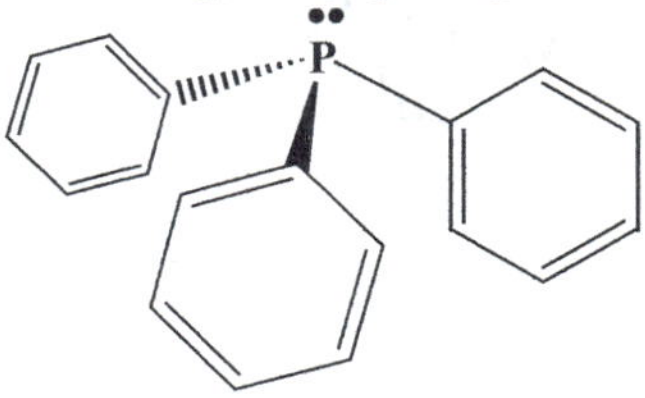

**trimethylphosphite**

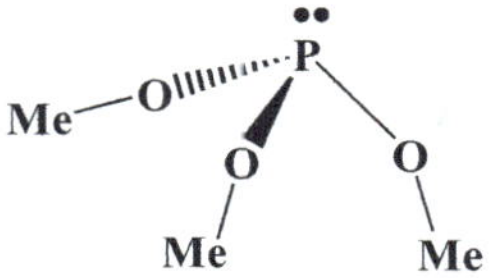

**triphenylphosphite**

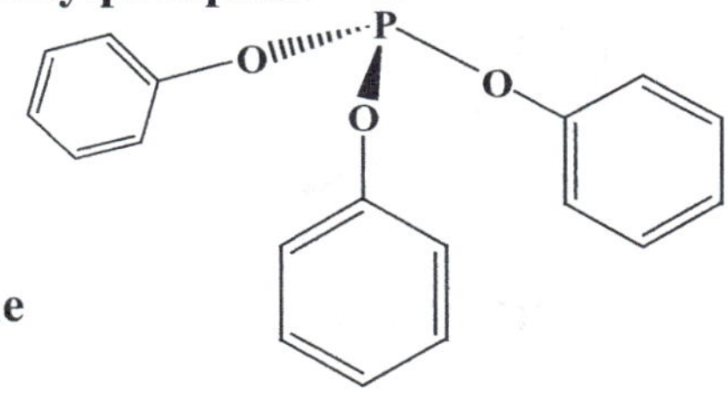

**tricyclohexylphosphine**

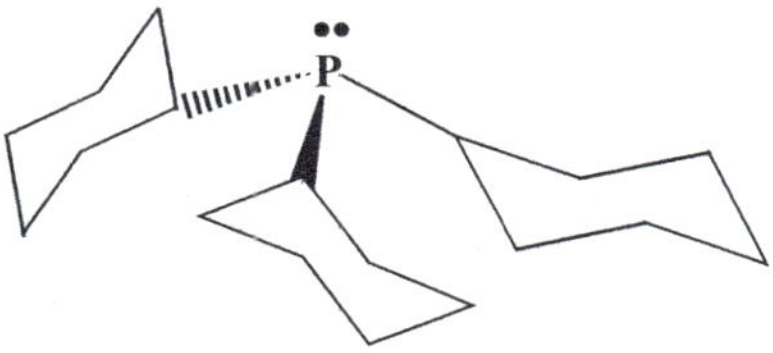

**trifluorophosphine**

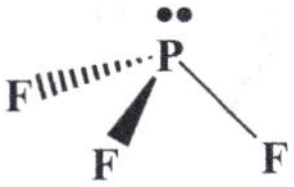

## Bidentate Organophosphine Ligands

**dmpm**

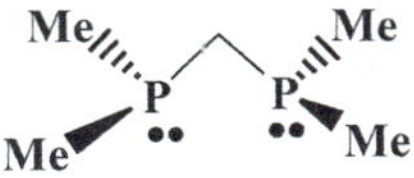

**dmpe**

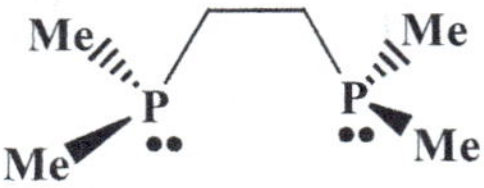

**dppm**

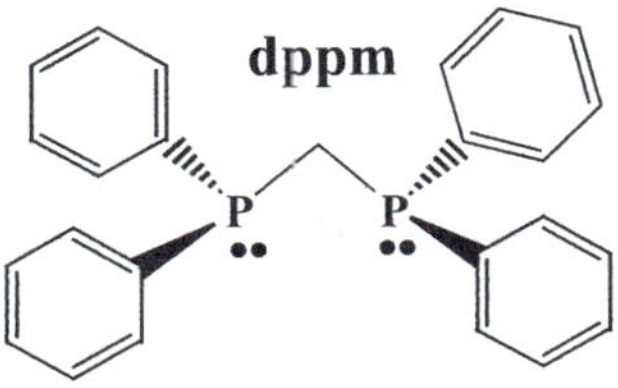

**dppe**

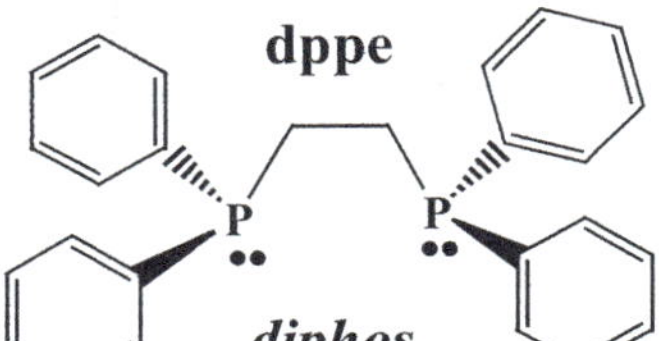

*diphos*

## Chiral Organophosphine Ligands

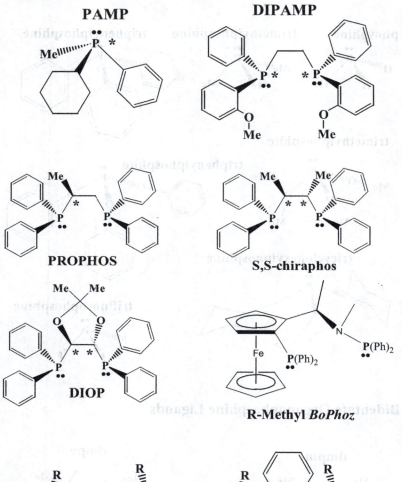

* indicates a chiral center

## Nomenclature of Coordination Compounds

The nomenclature of simple coordination compounds is developed in a set of rules for referring to ionic and neutral ligands, the number of each type of ligand, and the oxidation state of the metal. A number of examples of naming compounds and writing formulas are given.

The nomenclature of coordination compounds is introduced in two sections. First, we consider the basics of naming ligands (including multidentate, ambidentate, and bridging) that occur in simple neutral as well as ionic coordination compounds, and then secondly we will consider naming the coordination compounds themselves.

Table 3.2 lists six rules for naming ligands and simple coordination compounds.

The name of anionic ligands is modified by removing the -*ide* suffix of halides, oxides, and hydroxides and replacing in these ligands with -*o*. Therefore, chloride becomes chloro, oxide becomes oxo, hydroxide becomes hydroxo, and so forth.

If the ligand has an ending such as -*ate* or -*ite*, the last *e* is removed and replaced with -*o*, so that nitrate becomes nitrato, sulfite becomes sulfite. But, there are a few exceptions to this rule. For example, amide becomes amido and the N-bonding form of the ambidentate nitrite becomes nitro. The very few positive ligands that are found in coordination compounds are modified by adding an –*ium* suffix to the root name.

The names of neutral compounds that act as ligands are usually not modified; however, it is customary to give a few common neutral ligands special names. These special cases are: water becomes aqua, ammonia is called ammine, carbon monoxide is carbonyl, and nitrogen oxide is nitrosyl. Molecular nitrogen and oxygen are referred to as dinitrogen and dioxygen.

In naming a coordination compound, the cation is named first and the anion is named second. This is the same method used to name ordinary inorganic salts, like sodium bromide or ammonium phosphate.

Next, in naming a coordination compound, the ligands are always named first (in alphabetical order), followed by the name of the metal. (This can lead to ambiguity in naming compounds: you might ask which comes first---chloro or dibromo? The answer is that you use the name of the ligand in determining order---not the prefix denoting how many of them are present in the complex. Thus: di*bromo* and then *chloro* when naming.)

One should note that the opposite order is followed in writing the formulas for coordination compounds; the symbol for the metal precedes the formulas for the ligands. An example is the compound tetraamminedichlorocobalt(III) chloride, where the written formula is: [Co(NH$_3$)$_4$Cl$_2$]Cl.

As shown in the example above, the oxidation state of the metal is indicated by use of Roman numerals in parentheses after the name. Therefore an oxidation state of zero is indicated by a numeral zero, 0, in the parentheses, an oxidation state of +1 is (I), +2 is (II), +3 is (III), and so forth.

If the metal containing portion of the coordination compound is an anion, the –*ate* suffix is added to the name of the metal, but the –*ium* or other suffix has to be removed from the name of the metal before the –*ate* is added. For example, titanium becomes titanate, and the other metals such as manganese becomes manganate, and molybdenum becomes molybdenate. Some metals named with a Latin root, such as iron, copper, silver, and gold, will retain the Latin stem for the metal and become ferrate, cuprate, argentate, and aurate when they are anions. The number of ligands is indicated by the appropriate prefix given in Table 3.2.

There are two sets of prefixes, one for monoatomic ions and polyatomic ions with relatively short names (*di-*, *tri-*, *tetra-*, etc.), and a second set of prefixes (*bis-*, *tris-*, *tetrakis-*, etc.) for ligands that already contain a prefix—for example, ethylenediamine or triphenylphosphine—or for complicated ligands whose names commonly appear in parentheses.

The use of parentheses is not as systematic in practice as might be expected, and is often used to help in clarification between different ligands, so that they won't be confused, and to help in setting off subscripts when the ligand itself has a subscript in the formula. Also, ionic ligands with particularly long names and neutral ligands without special names are enclosed in parentheses. So, for example, EDTA, or ethylenediammine-tetraacetato, is generally enclosed in parentheses, whereas cyano is not.

There are two different ways to describe ambidentate ligands. The first way is to put the symbol of the donating atom before the name of the ligand. This indicates which atom is coordinating to the metal, and is the preferred way to describe ambidentate ligands in that it infers the different ways the ligand could bind to the metal. Thus, –NCS would be N-thiocyanato, and –SCN would be called S-thiocyanato. The second way is to use a slightly different form of the name, which is dependent on the atom that is donating the electron pair to the metal. Thus, –NCS would be isothiocyanato, and –SCN would be called thiocyanato.

Bridging ligands are named by placing the Greek letter μ (mu) before the name of the ligand. Therefore, a bridging amide ($NH_2^-$), a bridging hydroxide ($OH^-$), or a bridging sulfide ($S^{2-}$) ligand becomes μ-amido, μ-hydroxo, or μ-sulfido, respectively.

If there is more than one of a given bridging ligand, the prefix indicating the number of ligands is placed after the μ and before the ligand. Thus, if there are two bridging hydroxo ligands they are indicated by using *μ-dihydroxo*.

If there are two bridging ligands, one bromide and one chloride for example, they are given in alphabetical order as *μ-bromo, μ-chloro*.

## Table 3.2. Nomenclature Rules for Simple Ligands

1. Anionic ligands end in –o.

| | | | | | |
|---|---|---|---|---|---|
| $F^-$ | fluoro | $SO_3^{2-}$ | sulfito | $OH^-$ | hydroxo |
| $Cl^-$ | chloro | $SO_4^{2-}$ | sulfato | $CN^-$ | cyano |
| $Br^-$ | bromo | $S_2O_3^{2-}$ | thiosulfato | $NC^-$ | isocyano |
| $I^-$ | iodo | $ClO_3^-$ | chlorato | $SCN^-$ | thiocyanato |
| $O^{2-}$ | oxo | $CH_3COO^-$ | acetato | $NCS^-$ | isothiocyanato |

2. Neutral ligands are named as the neutral molecule.

| | |
|---|---|
| $C_2H_4$ | ethylene |
| $(C_6H_5)_3P$ | triphenylphosphine |
| $NH_2CH_2CH_2NH_2$ | ethylenediamine |
| $CH_3NH_2$ | methylamine |

3. There are special names for four neutral ligands:

$H_2O$ is *aqua*, $NH_3$ is *ammine*, CO is *carbonyl*, NO is *nitrosyl*.

4. Cationic ligands end in *–ium*.

$NH_2NH_3^+$   hydrazinium

5. Ambidentate ligands are indicated by:
   a. Placing the symbol of the coordinating atom in front of the name of the ligand, for example, *S*-thiocyanato and *N*-thiocyanato for –SCN- and –NCS

   b. Using special names for the two forms, for example, nitro and nitrito for $-NO_2^-$.

6. Bridging ligands are indicated by placing a µ-before the name of the ligand.

## Table 3.3. Naming Simple Coordination Compounds

1. As in simple salts, the name of the cation is given first.

2. The contents of the coordination sphere are specified: first the ligands, then the metal.

(a) The nature of the ligands is indicated by a suffix: positively charged ligands have the suffix *-ium*, neutral ligands no suffix, and negative ligands usually have the suffix *-o* (but $H_2O$ and $NH_3$ are called *aquo* or *aqua* and *ammine*, respectively). **(See Table 3.2)**

(b) The ligands are named in a definite order according to their charge: anionic ones first, then neutral, then cationic. Conventionally, but not necessarily conveniently, the names are not separated by spaces or hyphens.
**(See Table 3.2)**

(c) The number (two, three, four) of each kind of ligand is indicated by prefixes, *di, tri, tetra-* (for ligands having simple names such as chloro), or *bis-, tris, or tetrakis-* (for ligands having complicated names such as 2,4-pentanediono). **(See Table 3.4)**

(d) Finally, the name of the metal is given, the oxidation state is indicated by a Roman numeral in parentheses, and the ionic nature of the complex is indicated by the suffix *-ate* for an anionic complex or no suffix for cationic or neutral complexes.

## Table 3.4. EXAMPLES OF NOMENCLATURE OF COORDINATION COMPOUNDS

$[PtCl_2(NH_3)_4]Br_2$ = Dichlorotetraammineplatinum (IV) bromide
$[Al(OH)_2(H_2O)_4]^+$ = Dihydroxotetraaquoaluminum (III) ion
$[Co(NO_2)(NH_3)_5]SO_4$ = Nitropentaamminecobalt (III) sulfate
$[Co(ONO)(NH_3)_5]SO_4$ = Nitritopentaamminecobalt (III) sulfate
$[Co(H_2NCH_2CH_2NH_2)_3]Cl_3$ =
    Tris(ethylenediamine) cobalt (III) chloride
$NH_4[Cr(NCS)_4(NH_3)_2]$ =
    Ammonium tetra-n-thiocyanatodiammine-chromate (III)
$Na_3[Fe(CN)_6]$ = Sodium hexacyanoferrate (III)
$Na_3[Co(NO_2)_6]$ = Sodium hexanitrocobaltate (III)
$[Zr(acac)_4]$ = Tetrakis(2,4-pentanediono) zirconium (IV) or tetrakisacetylacetonato zirconium(IV)

### Examples:

The best way to understand Table 3.1, Table 3.2, Table 3.3, Table 3.4, and the text explanation is by working through a series of examples.

The rules for writing these formulas, as for all chemical compounds, are determined by the International Union of Pure and Applied Chemistry (IUPAC). We first consider naming compounds for which we are given a formula.

### Example A

Name the compound $[Co(NH_3)_4Br_2]Br$.

One starts first by naming the complex cation. The ligands are named alphabetically with ammine first and then bromo, and since there are four ammonias and two bromides the prefixes *tetra-* and *di-* are used. The cobalt oxidation state is determined as follows: The net charge on the complex cation must be $1^+$ to balance the one $1^-$ bromide anion. Since there are two $1^-$ bromides in the coordination sphere, the cobalt must be +3 oxidation state in order for the net charge on the cation to come out as $1^+$. Thus, the full name of the compound is:

tetraamminedibromocobalt(III) chloride
or   dibromotetraamminecobalt(III) chloride

**Example B**
Name the compound $(NH_4)_2[Pd(SCN)_6]$.

Here is a palladium-containing complex (which is an anion) and the counter ion is the common ammonium ion, $NH_4^+$, as the cation. Since the ligand is written with the S symbol first, it is the thiocyanato (or, alternatively, *S*-thiocyanato) form of the ambidentate ligand. There are six of these thiocyanate ligands, so one must use the *hexa*-prefix. The metal complex anion must have a net charge of $2^-$ to balance the two $1^+$ ammonium cations. Since the thiocyanate ion is also $1^-$, the palladium must be in the +4 oxidation state to give a net $2^-$ charge to the anion. Because the palladium is contained in a complex anion, its *–ium* suffix is removed and replaced with *–ate*. Therefore the full name of the compound is:

ammonium hexathiocyanatopalladate(IV)

**Example C**
Name the compound $[Zn(en)_2]SO_4$.

The zinc complex is a complex cation since there is a sulfate counter ion. There are two ethylenediamine ligands, which are shortened to *en*, but since en is a neutral ligand with *di-* already within its name, the prefix *bis-* is used. The oxidation state of the zinc will be the same as the charge on the overall complex cation since the ligands are neutral. The charge on the metal complex has to be $2^+$ to balance the $2^-$ of the sulfate ion, so the zinc is in the +2 oxidation state. The full name of this compound is therefore:

bis(ethylenediamine)zinc(II) sulfate

Inorganic Chemistry: Introduction to Coordination Chemistry

**Example D**
Name the compound [Ag(CH$_3$NH$_2$)$_2$]$_2$SO$_4$.
  One must use the *bis-* prefix for the two neutral methylamine ligands. The silver metal complex is a cation, and there are two of them so as to balance out the 2⁻ charge on the sulfate ion. Thus, the oxidation state of the silver has to be +1. The full name of this compound is therefore:

bis(methylamine)silver(I) sulfate

**Example E**
Name the compound N(butyl)$_4$[Re(H$_2$O)$_2$(C$_2$O$_4$)$_2$].
  The tetrabutyl ammonium ion, and all ammonium type ions, are 1⁺ charged, thus the rhenium metal complex is an anion with a 1⁻ charge. In the metal complex anion, the two neutral aqua (water) ligands come alphabetically before the two anionic oxalate ligands. The oxalate ligands are each 2⁻ charged. Since the two oxalate anionic ligands contribute 4⁻ charges, and the overall complex has a 1⁻ charge, then the rhenium is in the +3 oxidation state. The name rhenium is amended to rhenate because this metal is in a complex anion. Therefore the full name of this compound is:
Tetrabutylammonium diaquadioxalatorhenate(III)

**Example F**
Name this compound:

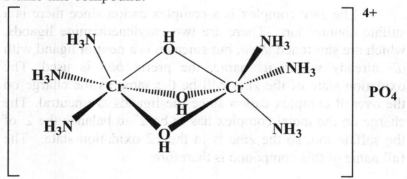

This is an example of a bridged compound. The three hydroxide ligands bridge between the two chromium ions.

One names such compounds from left to right remembering to place a μ in front of the bridging ligands. The oxidation states of the two chromium metal atoms could be (III) and (III), (II) and (IV), (I) and (V), or any other combination adding up to 6. If one were to find out that the two metal atoms have the same oxidation state, then the oxidation state of chromium would be +3. The full name of the compound is therefore:

triamminechromium (III)- μ-trihydroxo triamminechromium (III) phosphate

Instead of writing the formula from the structure, it is instructional to work a few examples in which one is given the names of some coordination compounds and then to supply the correct formulas or structures.

The IUPAC rules concerning the order in which the formulas of the ligands in a coordination compound should be written are complicated and generally not treated in an introductory textbook. We will follow the common, but not officially correct method, practice of writing the ligand formulas of a coordination compound in the same order they are named, which is in alphabetical order by the first letter of the name of the ligand.

**Example G**

Write the formula for the compound bis(acetylacetonato)diaquarhodium(III) chloride.

The formula for the bidentate acetylacetonate ligand is given in Table 3.1, but this 1⁻ anion is usually abbreviated as **acac**. The acac and two water molecules that constitute the coordination sphere along with the rhodium(III) cation are set apart in brackets. The net charge on the complex cation is $1^+$ (because the acac is 1⁻ and there are two of them); so one chloride counter anion is needed. The formula of the compound is therefore:

$[Rh(acac)_2(H_2O)_2]Cl$

# Inorganic Chemistry: Introduction to Coordination Chemistry

**Example H**
Write the formula for the compound diamminebromo(ethylene)nitrotriphenylphosphineplatinum (IV) phosphate.

This compound has five different types of ligands in the coordination sphere: $NH_3$, $Br^-$, $C_2H_4$, $NO_2^-$, and $PPh_3$. The real difficulty in constructing this formula is figuring out how many cations and anions there must be. The cation has a net charge of $2^+$, and the anion is $3^-$. Therefore, there must be three cations and two anions to ensure electrical neutrality. The formula for this compound is therefore:

$$[Pt(NH_3)_2Br(C_2H_4)NO_2PPh_3]_3(PO_4)_2$$

**Example I**
Draw the structural formula for:
tetraamminemolybdenum (III)- μ-chloro- μ-hydroxo bis (ethylenediamine)ruthenium(III) sulfate.

The $OH^-$ and $Cl^-$ ions are bridging ligands between the cobalt and iron cations. The overall charge on this huge cation is $4^+$ because there are $6^+$ charges from the two $3^+$ cations and $2^-$ charges from the two $1^-$ bridging hydroxo and chloro anions. Therefore, there must be two ($2^-$) sulfates in the formula:

# Inorganic Chemistry: Introduction to Coordination Chemistry

## Terms and Definitions  Chapter 3:

*Ligand*
*Hard and Soft*
*Polarizable*
*Nucleophiles*
*Electrophiles*
*Coordinate Covalent Bond*
*Classical Or Simple Donor Ligands*
*Nonclassical Ligands*
*π-Bonding or π-Acid –Ligands*
*One-Electron Donors*
*Two-Electron Donor*
*Three-Electron Donors*

*Unidentate*
*Bidentate*

*Chelate*

*Bridging Groups*

*Bidentate Bridging*
*Multidentate Ligands*

*Inherent Thermodynamic Stability*

*Linkage Isomerism*
*Ambidentate Ligands*

*Heterobimetallic*
*Homobimetallic*

*Macrocyclic*
*Overall Formation Constant*
*Overall Stability Constant*

# Inorganic Chemistry: Introduction to Coordination Chemistry

**Chapter 3 Problems and Exercises**

1. In the compound $Al_2Cl_6$, two of the chloro ligands are *bridging* and four are terminal. Draw this structure.

2. Describe whether the following ligands are monodentate or bidentate:
(A) ethylenediamine
(B) aqua
(C) cyano
(D) carbonyl
(E) ammine

3. In Lewis acid base theory, a d-metal complex consists of a central metal atom to which are attached a number of ligands. The metal is a _____ and the ligands are _____. (Lewis Acid or a Lewis Base?)

4. Which of the following ligands are considered to be one-electron donors?
 (a) chloride  (b) ammonia  (c) water  (d) $CH_3^-$  (e) iodide

5. Is the $Cu^{2+}$ ion considered to be a Lewis acid, or a Lewis base?

6. Which of the following ligands in each listed pair is considered to be the softest?
 (a) $F^-$ or $I^-$
 (b) $S^{2-}$ or $O^{2-}$
 (c) $PH_3$ or $AsH_3$
 (d) en or dppe
 (e) thiocyano or isothiocyano
 (f) EDTA or dtc (dithiocarbamate)
 (g) ammonia or pamp

7. Which metal in each pair listed is the hardest?
 (a) $Cu^+$ or $Cu^{2+}$
 (b) $Pd^{2+}$ or $Ni^{2+}$
 (c) $Cr^{3+}$ or $Au^{3+}$

# Inorganic Chemistry: Introduction to Coordination Chemistry

8. Which of the following ligands in each pair listed would *most likely* have the greatest inherent stability in binding to a metal ion?
   (a) Trimethyl phosphite or diphos
   (b) ammonia or EDTA
   (c) acac or water
   (d) chloride or diop
   (e) en or porphyrin
   (f) dmpe or s,s-chiraphos
   (g) trien or cyclam

9. Label each of the following ligands as simple classical ligands or non-classical ligands.
   (a) ammonia (b) water (c) chloride (d) EDTA (e) en
   (f) diphos (g) acac (h) DuPhos (i) porphyrin
   (j) dipamp (k) carbon monoxide (l) methylamine

10. Draw the structures of the two compounds formed when dmpe and dmpm both are used as a chelating bidentate ligand in separate reactions with chromium hexacarbonyl, $Cr(CO)_6$. What is the difference in the chelate rings formed by these two ligands in the reaction with $Cr(CO)_6$

11. Name the following compounds:
    (a) $[Pt(NH_3)_4Cl_2]SO_4$
    (b) $K_3[Mo(CN)_6F_2]$
    (c) $K[Co(EDTA)]$
    (d) $[Co(NH_3)_3(NO_2)_3]$

12. Name the following compounds:
    (a) $[Pt(NH_3)_6]Cl_4$
    (b) $[Ni(acac)(P(C_6H_5)_3)_4]NO_3$
    (c) $(NH_4)_4[Fe(ox)_3]$
    (d) $W(CO)_3(NO)_2$

13. Name the following compounds:
    (a) $[Pt\{P(C_6H_5)_3\}_4](CH_3COO)_4$
    (b) $Ca_3[Ag(S_2O_3)_2]_2$
    (c) $Ru(As(C_5H_5)_3)_3Br_2$

14. Name the following compounds:
(a) $[Fe(en)_3][IrCl_6]$
(b) $[Ag(NH_3)(CH_3NH_2)]_2[PtCl_2(ONO)_2]$
(c) $[VCl_2(en)_2]_4[Fe(CN)_6]$

15. Write formulas for the following compounds:
(a) Pentaammine(dinitrogen)ruthenium(II) chloride
(b) Aquabis(ethylenediamine)thiocyanatocobalt(III) nitrate
(c) Sodium hexaisocyanochromate(III)

16. Write formulas for the following compounds:
(a) Bis(methylamine)silver(I) acetate
(b) Barium dibromodioxalatocobaltate(III)
(c) Carbonyltris(triphenylphosphine)nickel(0)

17. Write formulas for the following compounds:
(a) Tetrakis(pyridine)bis(triphenylarsine)cobalt(III) chloride
(b) Ammonium dicarbonylnitrosylcobaltate(-I).
(c) Potassium octacyanomolybdenate(V)
(d) Diamminedichloroplatinum(II)

18. Write formulas for the following compounds:
(a) Hexamminecobalt(III) pentachlorocuprate(II)
(b) Tetrakis(pyridine)platinum(II) tetrachloroplatinate(II)
(c) Diammine bis(triphenylphosphine) palladium(II)
(d) bis(oxalato) aurate(III)

19. Draw a valid structure for each of the following compounds:
(a) (Ethylenediamine)iodonitritochromium(III)- µ-dihydroxotriamminechlorocobalt(III) sulfate
(b) Bis(ethylenediamine) cobalt(III)- µ-isocyano- µ-thiocyanatobis(acetylacetonato) chromium(III)nitrate

# Inorganic Chemistry: Introduction to Coordination Chemistry

20. Draw a valid structure for each of the following compounds:
(a) Pentamminechromium(III)- μ-hydroxopentamminechromium(III) chloride
(b) Diammine(ethylenediamine)chromium(III)- μ-bis(dioxygen)tetraaminecobalt(III) bromide.

21. The ligand thiocyanate (SCN-) is an ambidentate ligand. Draw two valid tetrahedral metal complexes of $Zn^{2+}$ of this ligand that each exhibit one of the ambidentate bonding modes that this ligand can use to bind to the metal.

22. Which of the following bidentate chelating ligands would probably exhibit the most steric hindrance around a metal ion?
   (a) dmpm or dmpe
   (b) dppe or dppm
   (c) dmpe or dppe
   (d) dppe or dipamp
   (e) triphenyl phosphine or triphenyl phosphite
   (f) trien or 2,3,2-tet

23. Label the following compounds as homobimetallic or heterobimetallic.
   (a) $Co_4(CO)_{12}$  (b) $Co_3Rh(CO)_{12}$
   (c) $Mn_2(CO)_{10}$  (d) $Fe_2Ru(CO)_{12}$

24. The total displacement reaction of water with ammonia ligand, on aqueous cobalt 3+ is given as:

$[Co(H_2O)_6]^{3+} + 6\,NH_3 \leftrightarrow [Co(NH_3)_6]^{3+} + 6\,H_2O$

If the stepwise stability constants for the reaction at a certain temperature and pH is:
Log $K_1$ = 2.79, Log $K_2$ = 2.26, Log $K_3$ = 1.69, Log $K_4$ = 1.25, Log $K_5$ = 0.74, Log $K_6$ = 0.03.

(a) calculate log $\beta_6$ for $[Co(NH_3)_6]^{3+}$
(b) calculate $\beta_6$

# Chapter 4. Stereochemistry of Coordination Compounds

Coordination compounds exhibit a wide variety of isomers due to large coordination numbers (usually 4-6) and the differing ways in which ligands can bind to the metal center.

There are many types of isomers in coordination chemistry, for example: *solvent isomers, ionization isomers, coordination isomers*; all of these have different ligands in the coordination sphere, with the same overall formula. The names of these *constitutional isomers* indicate whether solvent, anions or other coordination compounds form the changeable part of the structure.

*Stereoisomers* are isomers with the same ligands but with different shapes. These *configurational isomers* are the main focus of this chapter.

The following diagram may help make the distinction between isomer types clearer.

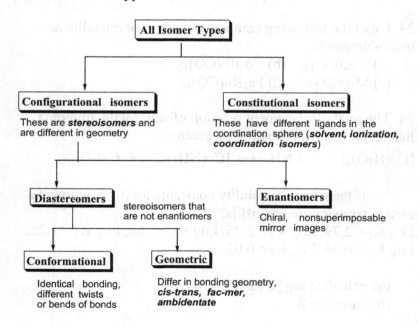

The terms *cis and trans* are common in organic chemistry and have approximately the same meaning for coordination compounds. However, the terms *fac* and *mer* are unique to coordination chemistry, and like the terms *cis* and *trans*, they describe different bonding geometries of ligands around the metal center. The terms **linkage isomerism** or **ambidentate isomerism** are used for cases of bonding through different atoms of the same ligand as discussed in the previous chapter.

**Configurational Isomers**

Examples of typical stereoisomerism are discussed next for four coordinate metal complexes (square planar) and six coordinate metal complexes (octahedral). Four coordinate geometries common for transition metals are square planar and tetrahedral. The tetrahedral geometry is so thoroughly covered in organic classes, with so many examples, that it is really not appropriate to cover that material in this text. On the other hand, square planar and octahedral geometries are not covered in organic classes.

**Square Planar**

Square planar complexes may exhibit geometrical isomers when there is more than one ligand. There can be up to four separate ligands on a metal with a square planar geometry. If A, B, C, and D are the four different ligands then we can have the following metal compounds with geometrical isomers: $(MA_2B_2)$, $(MA_2BC)$, and $(MABCD)$.

The $MA_3B$ compound (or any of its analogs with ligands C or D) would only have one isomer.

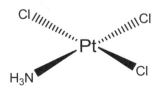

The examples for the three cases are discussed below.

**4-coordinate square planar ($MA_2B_2$) = two isomers**
Cis and trans isomers of $[PtCl_2(NH_3)_2]$

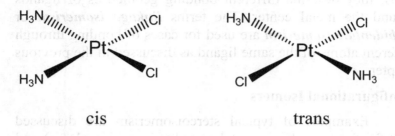

cis                         trans

**4-coordinate square planar ($MA_2BC$) = two isomers**
Just as in the example above, there can only be two isomers:

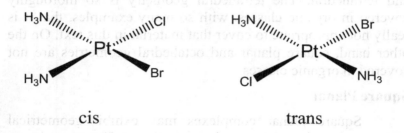

cis                         trans

**4-coordinate square planar (MABCD) = three isomers**

There are three isomers, and the key to seeing them is to notice what ligands are *trans* to each other. (Ammonia is trans to a different ligand in each of the three isomers)

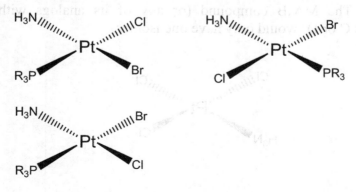

## Octahedral

The octahedral geometry is by its very nature much more complex than the square planar geometry when it comes to isomer determinations. It is best to look at the simplest cases first.

The simplest case of isomer formation for an octahedral complex is when there is only one kind of ligand, such as ethylene diamine, a bidentate chelating ligand. For the case where there are three bidentate chelating ligands around the metal center, such as in $[Co(en)_3]^{3+}$, there is the possibility of two *enantiomers* (enantiomers are isomers that are *non-superimposable mirror images*).

These two stereoisomers are shown in the next scheme. They can be thought of as propeller blades to give a better feel for their structure. These are *mirror images*.

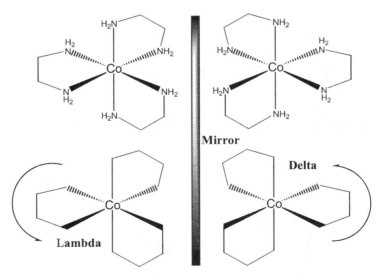

Molecules like these that are *nonsuperimposable mirror images* of each other are said to be *chiral*.

To decide if a particular isomer has a Lambda or Delta configuration one must look down the 3-fold rotation axis and see if the propeller rotates counterclockwise (Lambda, Δ) or clockwise (Delta, Λ).

For an octahedral complex with two different ligands there can be various types of ligands, such as monodentate, bidentate, or even tridentate. With a tridentate ligand such as dien, in the following case **[M(L-L-L)X$_3$]**, two geometric isomers are possible:

[Co(dien)Cl$_3$] with a ***tridentate chelating ligand***

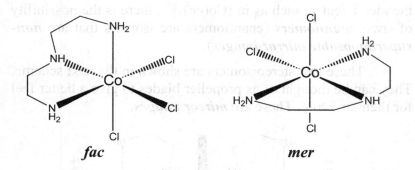

*fac*          *mer*

The fac isomer (facial) is named in that manner because it occupies one face of the octahedron, and the mer isomer (meridional) is named in that manner because it occupies the meridian of the octahedron (lies in a plane).

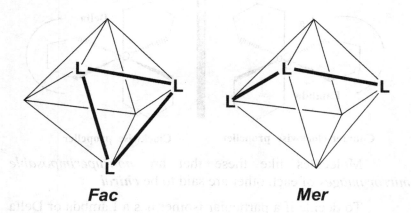

**Fac**          **Mer**

The situation on the previous page, where both fac and mer isomers are seen, is similar to the case where there are two monodentate ligands coordinated to the metal in equal numbers; only these same two geometric isomers are observed:

**[MA$_3$B$_3$] case** such as [Co(NH$_3$)$_3$Cl$_3$]

*fac* (facial)      *mer* (meridional)

Compare these fac and mer isomers to the ones on the preceding page.

With varying numbers of two different ligands specific isomer cases are observed. For example, only two geometric isomers are observed in the [MA$_4$B$_2$] case:

**[MA$_4$B$_2$] case** such as [Co(NH$_3$)$_4$Cl$_2$]$^+$

cis                    trans

Cis and trans isomers are seen in this case. Compare this set of isomers to the square planar case that was discussed previously. The similarity is that the ligands around the square plane determine the cis-trans isomer observed.

For monodentate ligands the following can be summarized:

**[MA$_5$B$_1$] case = one isomer only**
**[MA$_2$B$_4$] case = two isomers (cis and trans)**
**[MA$_3$B$_3$] case = two isomers (fac and mer)**
**[MA$_4$B$_2$] case = two isomers (cis and trans)**
**[MA$_1$B$_5$] case = one isomer only**

    A more complicated situation arises when there are two bidentate ligands and two monodentate ligands around the octahedral coordination sphere. In this case, not only do we observe geometric isomers (cis/trans), but one of the isomers also has an enantiomer (a non-superimposable mirror image).

An example is: **[M(L-L)$_2$X$_2$] case** such as [Co(en)$_2$Cl$_2$]$^+$

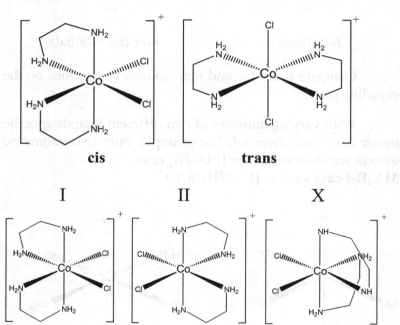

    For the cis geometric isomer, there are two enantiomers, these are mirror images I and II. Isomer II is not the same as structure X, which is shown so that a comparison can be made. Structures I and X are the same after simple rotation by 180° around the cobalt.

## [MA$_2$B$_2$C$_2$] case has six isomers (five are geometric)

The case where there are three different ligands, two of each, can be worked out systematically.

Ligands of the same type can either be cis or trans to each other in an octahedral complex. All three of the ligands can express a cis orientation, but only one ligand type at a time can be trans. If two ligand types are trans to each other, then all three must be trans. So, the following geometric isomers can be drawn for the [Co(NH$_3$)$_2$(Br)$_2$(Cl)$_2$]$^-$ ion:

(1) trans, trans, trans (2) cis, cis, cis
(3) trans, cis, cis (4) cis, trans, cis (5) cis, cis, trans

Thus, we have an all cis isomer, an all trans isomer, and three isomers that have two ligand types cis, and the other trans. The all cis isomer has an enantiomer. We can tell that the all cis isomer would have an enantiomer since it has no plane of symmetry through it, whereas we can draw a plane of symmetry through the other geometric isomers.

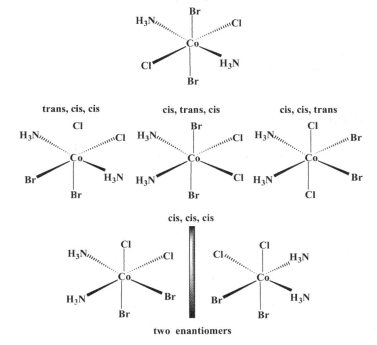

two enantiomers

For octahedral complexes with three or more different ligands the situation can become very complicated as far as total number of isomers, enantiomers, and equivalent structures. The best way to understand these is to work the exercises at the end of this chapter. These exercises use some real examples of compounds that exhibit numerous isomers.

**Constitutional Isomers**

In octahedral metal complexes the six octahedral positions are commonly numbered as below, with positions 1 and 6 in axial positions and 2-5 in counterclockwise order as viewed from the 1 position.

If six different monodentate ligands are completely scrambled (rather than limited to trans pairs as shown on the previous page), then there are 15 different diastereomers (different structures that are not mirror images of each other), each of which has an enantiomer (mirror image), which means there can be 30 different isomers! There is only one example of an octahedral Pt compound to date (that we are aware of) that has six different ligands around the metal.

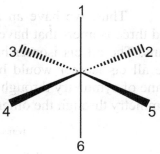

*Hydrate isomerism* is also called *solvent isomerism*.

The classical example is $CrCl_3 \times 6\ H_2O$. This compound can have three distinctive forms now known as:

    $[Cr(H_2O)_6]Cl_3$              violet
    $[CrCl(H_2O)_5]Cl_2 \times H_2O$    blue-green
    $[CrCl_2(H_2O)_4]Cl \times 2\ H_2O$    dark green

Draw these compounds, and which, if any, exhibit isomerism? Could you separate the three compounds above by using an ion exchange column? **Yes!**

What kind of resin would you use? Would you use an anion exchange resin, or a cation exchange resin?

***Ionization Isomerism*** is exhibited by compounds with the same formula, but which give different ions in solution. Examples are:

$[Co(NH_3)_5SO_4]NO_3$ and $[Co(NH_3)_5NO_3]SO_4$

$[Co(NH_3)_4NO_2Cl]Cl$ and $[Co(NH_3)_4Cl_2]NO_2$

***Coordination Isomerism*** *requires at least two metal ions.*
An example is the set of platinum compounds with the ***empirical formula*** of $[Pt(NH_3)_2Cl_2]$

There are three coordination isomers possible:
- $[Pt(NH_3)_4Cl][PtCl_4]$
- $[Pt(NH_3)_2Cl_2]$
- $[Pt(NH_3)_3Cl][Pt(NH_3)Cl_3]$

***Linkage Isomerism***:
Some ligands can bond to the metal through different atoms, and these are called ***ambidentate ligands***.

Examples:  thiocyanate SCN-   can bind through S or N
 nitrite $NO_2$-   can bind through N or O

**Chiral Ligands**

Some ligands are chiral in nature, and as in organic chemistry those ligands have central atoms, such as carbon, nitrogen, or phosphorous, that apparently have four different substituents around them.

In 1968 William S. Knowles changed the triphenyl phosphine ligands in Wilkinson's catalyst with the chiral ligand methyl-phenyl-propyl-phosphine. The modified catalyst was used in an asymmetric hydrogenation, and this lead to a slight excess of one of the enantiomer isomers (15% excess). This was a substantial increase, even though slight, and implied that this method of using chiral ligands could lead to total asymmetric hydrogenations.

Soon after that, the molecule L-Dopa was found to be useful in treatment of Parkinson's disease. Knowles immediately saw the benefit in using his new asymmetric catalysts to give Monsanto a huge edge in production. Within six months the synthesis of L-Dopa was at the 50 gallon per batch scale at a 94% enantiomeric excess using a rhodium catalyst with a chiral bisphosphine ligand.

The synthesis of organic specialty chemicals, and medicines like naproxen has increased dramatically in the last couple of decades. No longer does "mother nature" hold the only secret to asymmetric catalysts though the use of enzymes, humans have discovered the secret of ligand modification.

Chiral ligands confer certain shapes to metal complexes that they are bonded to, and that has also been found to be important in the area of nuclear medicine. One example is the technetium-99m radiopharmaceutical based on the ligand dimercapto-succinic acid, which has two chiral centers. When this ligand is reacted with Tc starting materials such as pertechnetate (TcO$_4^-$) or with the rhenium analog (ReO$_4^-$) then three different isomers are formed.

This Re compound is of interest to nuclear medicine clinicians because it (and the Tc analog) accumulates preferentially in some human tumors, especially medullary thyroid carcinoma. Since $^{99m}$Tc is a γ (gamma) emitting isotope, and the $^{188}$Re and $^{186}$Re isotopes are both primarily β (beta) emitting isotopes, then the possibility of a "matched pair " of diagnostic (Tc) and therapeutic (Re) agents for these cancers arises.

The ReO(DMSA)$_2^-$ compound is of interest to coordination chemists because of the possibility of more than one isomer being produced in the reaction.

These isomers are produced due to the fact that in this reaction two chiral bidentate chelating DMSA ligands bind to the Re=O metal center. The ligand DMSA is meso-dimercaptosuccinic acid. The meso conformation is due to the configuration of the two chiral centers on the ligand.

**SYN-ENDO**  **SYN-EXO**

**ANTI**

The three ReO(DMSA)$_2^-$ isomers can be identified by proton NMR and separated so that the radiopharmaceutical can be used. There are many examples in the literature of chiral ligands being used in asymmetric synthesis, or in biomedical applications.

Inorganic Chemistry: Introduction to Coordination Chemistry

Terms and Definitions for Chapter 4

*Solvent Isomers*

*Ionization Isomers*

*Coordination Isomers*

*Constitutional Isomers*

*Stereoisomers*

*Chiral Ligands*

*Configurational Isomers*

*Hydrate Isomerism*

*Solvent Isomerism*

*Coordination Isomerism*

*Empirical Formula*

*Enantiomeric Excess*

*Radiopharmaceutical*

## Chapter 4 Problems and Exercises

1. The octahedral compound [Co(NH$_3$)$_3$Br$_3$] exhibits *fac* and *mer* isomers.

    Draw these two isomers and label them *fac* and *mer*.

2. Ethylenediamine (en) can act as a bidentate chelating ligand.

    Draw the two structural isomers that can be formed for the complex [Ir(en)$_2$Cl$_2$]$^+$.

3. Draw all possible isomers of:
(a) octahedral [RuCl$_2$(NH$_3$)$_4$]
(b) square planar [IrH(CO)(PR$_3$)$_2$]
(c) tetrahedral [CoCl$_3$(H$_2$O)]
(d) octahedral [Co(en)(NH$_3$)$_2$Cl$_2$]$^+$

4. Give structural formulas, including all isomers, for the following:
(a) Diamminebromochloroplatinum(II)
(b) Diaquadiiododinitropalladium(IV)

5. Give structural formulas, including all isomers, for the following and label the isomers that are cis and trans*, *fac* and *mer***, and enantiomers***:
(a) [Pt(NH$_3$)$_3$Cl$_3$]$^+$  *
(b) [Co(NH$_3$)$_2$(H$_2$O)$_2$Cl$_2$]$^+$  ***
(c) [Co(NH$_3$)$_2$(H$_2$O)$_2$BrCl]$^+$  ***
(d) [Cr(H$_2$O)$_3$BrClI]$^+$  ***
(e) [Pt(en)$_2$Cl$_2$]$^{2+}$  *  ***
(f) [Co(NO$_2$)$_3$(dien)]  **

6. The compound [Co(NH$_3$)$_3$(H$_2$O)Cl$_2$]$^+$ has three isomers where the cis and trans isomers are *mer*, and the other isomer is *fac*. Draw these.

7. The ion [Co(NH$_3$)Br(en)$_2$]$^{2+}$ exhibits geometrical isomerism and optical isomerism. Draw these isomers and distinguish between them.

# Inorganic Chemistry: Introduction to Coordination Chemistry

8. How many geometric isomers are possible for the square planar complex [Pt(NH$_3$)(OH$_2$)(Cl)(Br)]? Draw them.

9. Determine whether the following complexes have a chiral metal center:
(a) [Ru(en)$_3$]$^{2+}$
(b) *fac*-[Rh(en)(H$_2$O)Cl$_3$]$^+$
(c) *cis*-[Ir(en)$_2$Br$_2$]$^+$

10. Use sketches to show that the square planar complexes (a) [Pt(NH$_3$)$_2$BrI] has only two isomers
(b) [Pt(py)(NH$_3$)BrCl] has three isomers

11. Werner prepared two compounds by reacting a solution of PtCl$_2$ with triethyl phosphine, P(C$_2$H$_5$)$_3$. After elemental analysis, the compounds analyzed for Pt 38.8%; Cl 14.1%; C 28.7%; P 12.4% and H 6.02%. Write formulas and structures for the two isomers.

12. What is the relationship between the following two linear complex ions?  [Cl-Ag-SCN]$^-$ and [SCN-Ag-Cl]$^-$

The complex ions are:

(a) linkage isomers  (b) coordination isomers (c) geometric isomers  (d) optical isomers (e) the same compound

13. Enantiomers are: ___
(a) superimposable mirror images with identical chemical formulae and the same chemical reactivities.
(b) nonsuperimposable mirror images with identical chemical formulae and the same chemical reactivities.
(c) nonsuperimposable mirror images with dissimilar chemical formulae but similar chemical reactivities.
(d) superimposable mirror images with identical chemical formulae and similar physical properties

14. What is the relationship between the following two complex ions?
A. [Co(NH$_3$)$_4$(NO$_2$)Cl]$^+$
B. [Co(NH$_3$)$_4$(ONO)Cl]$^+$

The complex ions are:
(a) coordination isomers. (b) optical isomers. (c) linkage isomers. (d) geometric isomers. (d) the same compound.

15. How many geometric isomers exist for the tetrahedral complex ion $[CoCl_2Br_2]^{2-}$?

16. Draw the octahedral cobalt ammonates that have their ligands in the following numbered spots in italics:
(a) Co(*1,2,5*-NH$_3$)$_3$(*3,4,6*-Cl)$_3$
(b) Co(*1,2,6*-NH$_3$)$_3$(*3,4,5*-Cl)$_3$

Which of these compounds is *fac* and which is *mer*?

17. In the octahedral cobalt compound $[Co(en)_2(Cl)_2]^+$, the isomers that has the ligands in the following numbered places [Co(*3,4*-en) (*2,5*-en) (Cl)$_2$]+ should be labeled cis, trans, fac, or mer?

18. Numerically label all the ligands for the enantiomers of $[Co(NH_3)_2(Br)_2(Cl)_2]^-$. Make the two chloride ligands occupy the 1,2 spots for each of these enantiomers.

19. Draw and numerically label all the ligands for the cis, cis, cis and the trans, trans, trans isomers of $[Co(NH_3)_2(Br)_2(Cl)_2]^-$.

Bonus Questions

20. What is the formula weight of $[Co(NH_3)_2(Br)_2(Cl)_2]_2SO_4$?

21. What is the formula weight of *fac*-[Rh(en)(H$_2$O)Cl$_3$]Br?

22. What is the formula weight of $(NH_4)_4[Mo(CN)_8]$ ?

23. If 150 mg of Mo(CO)$_6$ was reacted with DPPE how much product (Mo(CO)$_4$ DPPE) would be formed if the yield was only 55%?

23. If 1.55 mmol of the ReO(DMSA)$_2^-$ compound is isolated as its tetrabutyl ammonium salt, and the syn-endo isomer is synthesized at only 22% of the total, how many grams of the syn-endo (Nbutyl$_4$)[ ReO(DMSA)$_2$] was synthesized?

# Chapter 5.  Crystal Field Theory

Before discussing Crystal Field Theory it is instructive to note the deficiencies of Valence bond theory, and to examine electron counting formalisms.

**Valence-Bond Theory versus Crystal Field Theory**

Valence bond theory is built on the Lewis electron dot concept that covalent bonding results from sharing of electron pairs between atoms, and then put into a quantum-mechanical approach that was refined by Linus Pauling. The valence bond approach is very useful to organic chemists who use the theory very successfully, but it has been used for years by inorganic chemists---~~not~~ so successfully.

The main new concept of valence-bond theory is *the orbital hybridization concept*, which asserts that the wave functions of atomic orbitals of an atom (usually the central atom) can mix together during bond formation. The result is a *hybrid orbital*. The number of hybrid orbitals formed by this approach equals the number of atomic orbitals that are used to form the hybrid orbitals. Not only can s and p orbitals be utilized, *but d orbitals can be used* as well.

**Table 5.1  Hybridization of Central Atom Orbitals.**

| Hybridization | # of hybrid orbitals | Molecule geometry |
|---|---|---|
| $sp$       (1+1=2)      | 2 | Linear |
| $sp^2$     (1+2=3)      | 3 | Trigonal planar |
| $sp^3$     (1+3=4)      | 4 | Tetrahedral |
| $dsp^3$    (1+1+3=5)    | 5 | Trigonal bipyramidal |
| $d^2sp^3$  (1+2+3=6)    | 6 | **Octahedral** |

The formation of hybrid orbitals using valence-bond theory can be used to successfully explain and account for molecular shapes. Unfortunately however, there is <u>no actual evidence</u> for hybridization of orbitals, and the concept of hybridization of atomic orbitals is not a good predictive tool, and this is ~~especially true~~ for transition metal complexes.

Valence-bond theory cannot explain magnetic and color properties of transition metal complexes or the spectrochemical series of ligands.

**Electron Counting**

To be able to use Crystal Field Theory (CFT) successfully, it is essential that one can determine the electronic configuration of the central metal ion in any complex. This requires being able to recognize all the species making up the complex and knowing whether the ligands are neutral or anionic, so that one can determine the oxidation state of the metal ion. This is why we spent time looking at Lewis electron dot structures and ligands in the previous chapters.

In many cases the oxidation state for first row transition metal ions will be either (II) or (III), but in any case you may find it easier to start with the M(II) from which you can easily add or subtract electrons to get the final electronic configuration.

A simple and fun procedure exists for the M(II) case.

First write out all the first row transition metals with their symbols and atomic numbers:

| 22 | 23 | 24 | 25 | 26 | 27 | 28 | 29 |
|----|----|----|----|----|----|----|----|
| Ti | V  | Cr | Mn | Fe | Co | Ni | Cu |

To see the number of electrons in the 3d orbitals, ~~then cross off the first 2~~, hence:

| 2 | 3 | 4 | 5 | 6 | 7 | 8 | 9 |
|---|---|---|---|---|---|---|---|

So, the electronic configuration of **Ni(II) is $d^8$** and the electronic configuration of **Mn(II) is $d^5$**. What is the electronic configuration of Fe(III)? Well, using the above scheme, Fe(II) would be $d^6$, so by subtracting a further electron to make the ion more positive, the configuration of **Fe(III) will be $d^5$**.

This simple procedure works fine for first row transition metal ions, ~~but it doesn't for 2nd row elements.~~ (It is interesting to note however that the 3$^{rd}$ row transition elements starting with hafnium through to mercury do have the same last digits as the 2$^{nd}$ row transition elements!)

**Electron Counting Formalisms**

In the chapter on atomic structure in freshman chemistry we say that in the 4$^{th}$ period of the periodic table the orbitals are filled in the order [Ar]4s$^2$3d$^{10}$ for the transition metals. This turns out to be true only for isolated metal atoms that are not in a compound. If a metal ion is surrounded by an electronic field---by surrounding it with ligands---~~then the 3d-orbitals drop in energy and fill before the 4s orbitals~~.

Thus, an organometallic chemist naturally considers that the transition metal valence electrons are all d-electrons.

If we ask for the d-electron count on a transition metal such as Nb in the zero oxidation state, we call it d$^5$, not d$^3$. For zero-valent metals, we see that the electron count corresponds to the column it occupies in the periodic table. Therefore, Co is in the ninth column and is d$^9$ (not d$^7$) and Tc$^{3+}$ is d$^4$ (seventh column for Tc, and subtract three electrons). Since we can now assign a d-electron count to a metal center, we can determine the electronic contribution of the surrounding ligands in a metal complex and come up with an ~~overall~~ electron count.

**Metal Complex Electron Counting Method A:**

**The Ionic Charge Convention (ICC Method)**

The basic premise of this metal-complex electron counting method is that we first remove all of the ligands from the metal with the proper number of electrons for each ligand to bring it to a ***closed valence shell state***, which is usually an octet as we learned in Chapter 2.

For example, if we remove water from a metal complex, H₂O has a completed octet and acts as a neutral molecule. When it was bonded to the metal center it did so through its lone pair and there is no need to ~~change the oxidation state~~ of the metal to balance charge. We therefore call water a neutral two-electron donor, just as we did in Chapter 3.

By contrast however, if we remove a chlorine group from the metal and complete its octet, then we formally have chloride (Cl⁻). If we bond this chloride anion to the metal, the new lone pair forms our metal-chlorine bond and the <u>chloride ion acts as a two-electron ~~donor~~</u> ligand.

We must do the same thing with a ligand such as a methyl group -CH₃. In the case of a methyl group, the -CH₃ must come away from the metal as a methanide anion (CH₃⁻), so that the carbon can maintain an octet.

It is important to notice in these last examples using ~~anionic ligands~~ that to keep electrical charge neutrality we must formally oxidize the metal by ~~one~~ electron by assigning a positive charge to the metal. This ~~reduces~~ the d-electron count of the metal center ~~by one electron~~.

**Metal Complex Electron Counting Method B:**

**The Neutral Covalent Convention (<u>NCC</u> Method)**

The basic premise of this metal-complex electron counting method is that when we remove one of the ligands from the metal, rather than take the ligand to a closed shell state, we make it ***neutral***. Let's consider water once again. When we remove it from the metal, it is a neutral molecule with one lone pair of electrons. Therefore, as with the ionic charge convention method, water is a neutral two-electron donor.

We diverge from the ionic method when we consider a ligand such as chlorine or methyl. When we remove chlorine from the metal and make the chlorine fragment ~~neutral~~, we have a neutral chlorine **radical** (Cl·) rather than a chloride ion. Both the metal and the chlorine radical must donate one electron each to form the metal-ligand bond.

Therefore, the chlorine group is a one-electron donor, (as we stated back in Chapter 3. when we first looked at the ways in which to classify ligands) **not** a two-electron donor as it is under the ionic convention.

We must do the same thing with a ligand such as the methyl group -$CH_3$. In the case of a methyl group, the -$CH_3$ must come away from the metal as a methyl radical ($CH_3$·), so that the methyl group is electrically neutral.

In the neutral covalent convention, metals retain their full complement of d electrons because we never change the oxidation state from zero when we count electrons; i.e. Co will always count for 9 electrons regardless of the oxidation state and Nb will always count for five.

It is important to notice that this method does not give us any information about the formal oxidation state of the metal. Thus, we must go back and assign that in a separate step, and it is for this reason that ***many chemists prefer the ionic charge convention***. Chemists that are fluent in both methods of electron counting have a greater understanding of the true nature of metal-ligand bonding.

**Using the ICC Method to Calculate Oxidation State**

In a metal complex the oxidation state of the metal is calculated by determining the difference in the number of valence electrons in the zero oxidation state of the metal with the number of electrons present after the loss or removal of the ligands in their closed Lewis electron dot structures.

For example we can calculate the oxidation state of the metals for: $[MnO_4]^-$, $[PtF_6]$, $[Cr(NH_3)_6]^{3+}$, $[Cr(CO)_5]^{2-}$.

① For $[MnO_4]^-$ : $Mn^o$ has 7 e-'s,                 7
  overall charge on the complex is -1,    +1
  remove 8- charges for four $O^{2-}$     - 8
  *electrons left on the Mn*              0

The difference between the number of valence electrons on the metal (7) minus the number after removal of ligands (0): $(7) - (0) = +7$ is the oxidation state of Mn. Thus Mn(VII)

② For $[PtF_6]$ : $Pt^o$ has 10 e-'s,                 10
  overall charge on the complex is 0,     0
  remove 6- charges for six F-            - 6
  *electrons left on the Pt*              +4

The difference between the number of valence electrons on the metal (10) minus the number after removal of ligands (0): $(10) - (4) = +6$ is the oxidation state of Pt. Thus Pt(VI).

③ For $[Cr(NH_3)_6]^{3+}$ : $Cr^o$ has 6 e-'s,                 6
  overall charge on the complex is +3,    -3
  remove 0 charges for CO                 0
  *electrons left on the Cr*              +3

The difference between the number of valence electrons on the metal (6) minus the number after removal of ligands (3): $(6) - (3) = +3$ is the oxidation state of Cr. Thus Cr(III).

④ For $[Cr(CO)_5]^{2-}$ : $Cr^o$ has 6 e-'s,                 6
  overall charge on the complex is -2,    +2
  remove 0 charges for CO                 0
  *electrons left on the Cr*              +8

The difference between the number of valence electrons on the metal (6) minus the number after removal of ligands (8): $(6) - (8) = -2$ is the oxidation state of Cr. Thus $Cr^{-2}$.

## Crystal Field Theory

There are two things that set the study of the electronic structures of transition metal compounds apart from the body of valence theory.

One is the presence of partly filled d subshells. This leads to experimental observations not possible in most organic chemistry such as: paramagnetism, visible absorption spectra, and apparent irregular variations in thermodynamic and structural properties.

The second is that there is a crude but effective approximation, called *crystal field theory* that provides a powerful yet simple method of understanding and correlating all of those properties that arise primarily from the presence of the partly filled d subshells.

The crystal field theory provides a way of determining how the energies of the metal ion orbitals will be affected by the set of surrounding atoms or ligands by simple electrostatic considerations.

Crystal field theory is a *model* and not a realistic description of the forces actually at work, and is completely superseded by molecular orbital theory. With that being said, its simplicity and convenience have earned it a cherished place in the coordination chemist's "toolbox."

The electronic properties of transition metal complexes are discussed in terms of the "orbital splittings," which the crystal field theory enables us to work out relatively easily. Our attention will be confined entirely to the d-block elements, and will be focused primarily on those of the 3d series. This is where the crystal field theory works best. The splittings of f orbitals are generally so small that they are not chemically important.

In crystal field theory, it is assumed that the metal ions are **simple point charges** (cations with a positive charge) and the ligands are also simple point charges (anions with a negative charge). When applied to alkali metal ions containing a symmetrical sphere of charge, calculations of energies are generally quite successful. The approach taken uses classical potential energy equations that take into account the **attractive** and **repulsive** interactions between charged particles (that is, **Coulomb's Law interactions**).

Electrostatic Potential is proportional to $q_1 \times q_2/r^2$,

where $q_1$ and $q_2$ are the charges of the interacting ions and r is the distance separating them. This leads to the correct prediction that large cations of low charge, such as $K^+$ and $Na^+$, should **form few coordination compound**s.

For transition metal cations that contain varying numbers of d electrons in orbitals that are **NOT spherically symmetric**, however, the situation is quite different. The shape and occupation of these d-orbitals then becomes important in an accurate description of the bond energy and properties of the transition metal compound.

To be able to understand and use CFT then, it is absolutely essential to have a clear picture of the shapes of the d-orbitals.

One of the main differences between the **three *p*** orbitals and the **five *d* orbitals** is that the set of *p* orbitals has three identical orbitals oriented along the *x, y,* and *z* axes whereas the set of five *d* orbitals has four identical orbitals ($d_{xy}$, $d_{yz}$, $d_{xz}$, and $d_{x^2-y^2}$) and one (the $d_{z^2}$) that *looks* like it is special; that is, it *appears* to be rather different from the other four. This distinction between the $d_{z^2}$ orbital and the other four needs to be addressed in some detail, and follows after the next paragraph.

The following scheme illustrates the differences in the d-orbitals. Make sure that you understand and remember which d-orbitals are along which axes of the Cartesian coordinate system, and which d-orbitals are between the axes.

This is the most important distinction among the ~~five~~ d-orbitals, not their shapes!

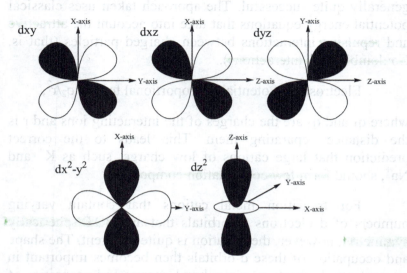

The Schrödinger mathematics, which yields the probability of finding electrons in various orbitals, can be carried out a number of different ways yielding a variety of solutions. *The set of five 3d orbitals depicted above is just one possible set of orbitals*.

Another solution of the Schrodinger mathematics in fact yields ~~six~~ dependent orbitals. These six dependent $d$ orbitals are the $d_{xy}$, $d_{yz}$, $d_{xz}$, $d_{x^2-y^2}$ orbitals shown above, plus ~~two~~ more similar to $d_{x^2-y^2}$, labeled $d_{z^2-y^2}$ and $d_{z^2-x^2}$.

The ~~five~~ 3$d$ orbitals that we normally use are generated from the six *dependent* orbitals by taking a ~~linear combination~~ of (that is by adding) the $d_{z^2-y^2}$ and $d_{z^2-x^2}$ orbitals to generate the $d_z^2$ orbital. When these two orbitals are added together, the resulting $d_z^2$ orbital has twice as great an electron probability along the z-axis as it does along the other two axes.

Keep in mind that the $d_z^2$ orbital is not as special as it looks; it is merely a ~~linear combination~~ of two dependent orbitals that look exactly like the other four orbitals in the $d$ subshell.

# Inorganic Chemistry: Introduction to Coordination Chemistry

The scheme below shows the five d-orbitals situated inside the Cartesian coordinate system (drawn with a hypothetical cube) at the origin with the x, y, and z axes labeled. Again, as discussed on the previous page, the $d_{x^2-y^2}$ and the $d_{z^2}$ orbitals lie along the axes, and the $d_{xy}$, $d_{yz}$, $d_{xz}$ orbitals do not.

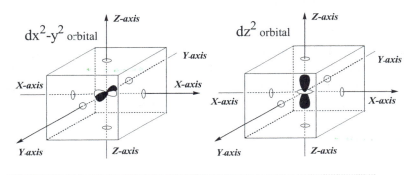

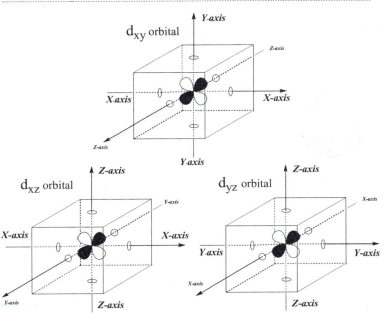

## Octahedral Geometry

Now, let us play a mind game and build a metal complex up from a free, gaseous metal ion with no ligands around it. The ~~five~~ d-orbitals in a gaseous metal ion are degenerate and have the same energy; this is illustrated in the following energy diagram.

If a spherical field of negative charges (a so-called crystal field) is placed around the metal ion, then the energy of the d-electrons in the d-orbitals is ~~increased~~ because of electrostatic repulsion (the d-electrons are negative charged just as the negative charges placed around the metal ion so they repel each other), but the five d-orbitals are still ~~degenerate.~~

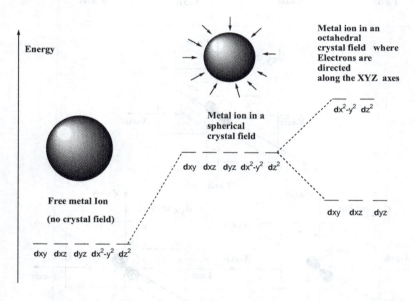

This is shown in the diagram above by an increased energy level, as compared to the free metal ion, for the ~~five d-orbitals~~, but they are all still degenerate.

Next, let us replace the spherical field of ~~negative~~ charge with a directed field. That is, let us move in six point charges, or ligands if you will, directly along the three axes of the ~~Cartesian coordinate~~ system, the X, Y, and Z-axes. This is the same geometry as an octahedral shaped metal complex, for instance $TiCl_6^{3-}$, or $[Co(NH_3)_6]^{3+}$.

In this octahedral case, things change dramatically. The five d-orbitals are no longer degenerate, and electrons in those d-orbitals that have lobes extending along the X, Y, and Z axes ($dx^2-y^2$ and $dz^2$) will experience greater repulsion due to these "incoming" ligands as compared to the electrons in the d-orbitals that have lobes that lie between the X, Y, and Z axes (*dxy, dyz* and *dxz*).

This is called *crystal field splitting*.

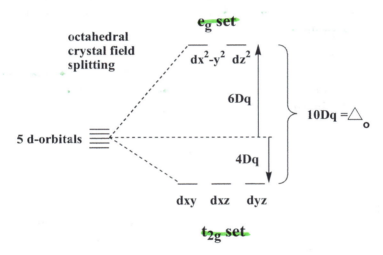

The result of these differences for the d-orbitals is an energy difference between the $dz^2$ and $dx^2-y^2$ orbitals compared to the *dxy, dyz* and *dxz* orbitals called **10Dq** by definition, or $\Delta_o$, which can vary in energy depending on metal and ligands, but is typically in the range of 100-300 kJ/mol.

This splitting energy is typified by an increase in energy for the two orbitals lying along the axes, called the *e_g set*, and a decrease in energy for the three orbitals lying in between the axes, called the *t_{2g} set*.

The $e_g$ set is *increased in energy by 6 Dq* from the "barycenter" and the $t_{2g}$ set is *decreased by 4 Dq*.

As mentioned above, the value of 10 Dq varies for different metal complexes.

Let's look at some metal complexes that exhibit these differences.

Consider the simple example of $TiCl_6^{3-}$ in which six chloride ions are in an octahedral arrangement around the $Ti^{3+}$ cation, which is $d^1$.

There is only one d-electron to be allocated to one of the five d- orbitals.

If it were to occupy the $dz^2$ or $dx^2-y^2$ orbital, both of which point directly towards the chloride ligands, it would be strongly repelled.

The geometry of the $dz^2$ or $dx^2-y^2$ orbitals and their nodes would require the electron to stay near the negatively charged ligands---causing even *more repulsion* than a spherically distributed electron would experience.

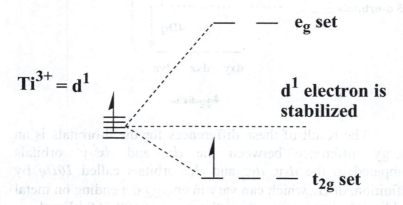

On the other hand, if the electron were to occupy the *dxy*, *dyz* or *dxz* orbital, it would spend less time near the ligands than would a spherically distributed electron and would be repelled less.

Thus, the $d^1$ electron is stabilized by 4Dq as explained by crystal field theory.

In the case described above for $TiCl_6^{3-}$ where there is only one d-electron, the electron goes into one of the $t_{2g}$ orbitals, and this has the effect of stabilizing the electron by -4 Dq. This stabilization by 4 Dq is called the *crystal field stabilization energy* (*CFSE*).

For a metal complex that has two d-electrons ($d^2$), then the CFSE would be twice as much, or -8Dq, (See HUND's RULE).

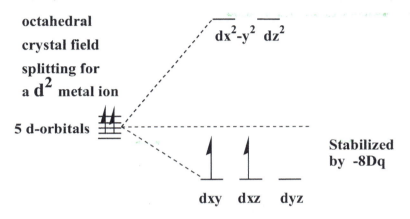

Electrons that are admitted into the $t_{2g}$ set are therefore stabilizing for octahedral metal complexes.

An octahedral metal complex with three d-electrons ($d^3$) would have a CFSE that is -12 Dq.

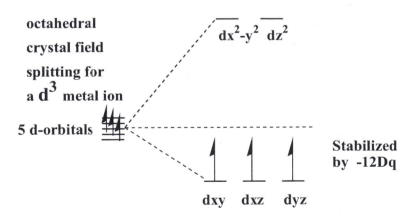

In the cases where there are four d-electrons ($d^4$) there are two possibilities that arise. *In the case where 10Dq is less than the pairing energy* for two electrons in the same orbital, the fourth electron will occupy one of the higher energy $e_g$ orbitals, and *in the case where 10Dq is greater than the pairing energy*, the fourth electron will pair up in one of the $t_{2g}$ orbitals. These are called the **weak field case** and the **strong field case** respectively. These cases are shown below with a relative illustration of the orbital energy splitting. The option for weak field or strong field is only available with $d^4$ through $d^7$ metal complexes.

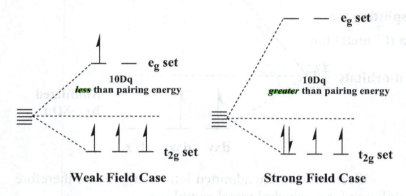

Weak Field Case         Strong Field Case

You should satisfy yourself that this statement is true by making a set of orbital energy diagrams like those above and fill in electrons for the situations where there are 8 to 10 d-electrons. For the cases shown above there are different numbers of paired and unpaired d-electrons. For the weak field case, there are four unpaired electrons, whereas in the strong field case there are only two unpaired electrons. The electron configurations can be written in this manner:

**Weak field**   $d^4 = \quad t_{2g}^3 e_g^1 \quad$ is also called *high spin*

**Strong field**   $d^4 = \quad t_{2g}^4 e_g^0 \quad$ is also called *low spin*

Therefore, the weak field case is also called *high spin*, and the strong field case is also called *low spin* because of the lower number of unpaired electrons.

The two possible cases where an octahedral metal complex can have a $d^5$ electron configuration are shown below:

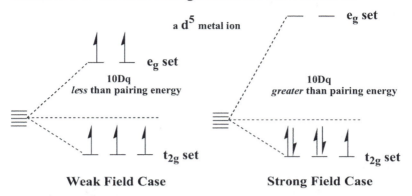

The $d^6$ and $d^7$ electron configurations are shown below:

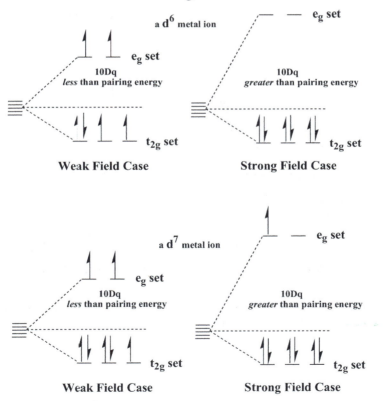

The $d^6$ strong field case has the maximum CFSE.

The $d^8$ and $d^9$ configurations show **NO DIFFERENCE** in their electron configurations in the weak and strong field cases; the electron configuration for $d^8$ is $t_{2g}^6 e_g^2$, and for $d^9$ it is $t_{2g}^6 e_g^3$.

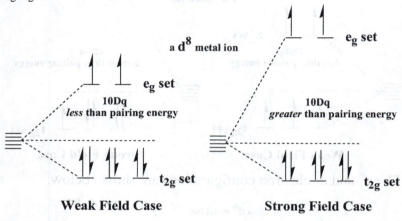

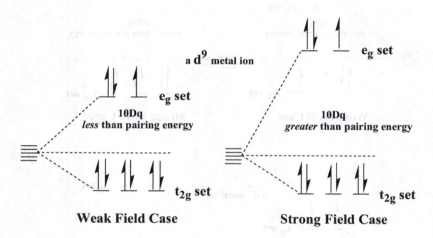

Of course, it goes without saying that the $d^{10}$ configuration is the ~~same~~ no matter what the field strength or ligand.

# Inorganic Chemistry: Introduction to Coordination Chemistry

The table below shows the crystal field effects for weak and strong octahedral fields. The P stands for the *pairing energy*.

**Table 5.2 Crystal Field Stabilization Energies for the Octahedral Geometry**

| WEAK FIELD | | | STRONG FIELD | | |
|---|---|---|---|---|---|
| dn | Config | CFSE | dn | Config | CFSE |
| $d^1$ | $t_{2g}^1 e_g^0$ | $-4Dq$ | $d^1$ | $t_{2g}^1 e_g^0$ | $-4Dq$ |
| $d^2$ | $t_{2g}^2 e_g^0$ | $-8Dq$ | $d^2$ | $t_{2g}^2 e_g^0$ | $-8Dq$ |
| $d^3$ | $t_{2g}^3 e_g^0$ | $-12Dq$ | $d^3$ | $t_{2g}^3 e_g^0$ | $-12Dq$ |
| | | | | | |
| $d^4$ | $t_{2g}^3 e_g^1$ | $-6Dq$ | $d^4$ | $t_{2g}^4 e_g^0$ | $-16Dq + P$ |
| $d^5$ | $t_{2g}^3 e_g^2$ | $0Dq$ | $d^5$ | $t_{2g}^5 e_g^0$ | $-20Dq + 2P$ |
| $d^6$ | $t_{2g}^4 e_g^2$ | $-4Dq + P$ | $d^6$ | $t_{2g}^6 e_g^0$ | $-24Dq + 3P$ |
| $d^7$ | $t_{2g}^5 e_g^2$ | $-8Dq + 2P$ | $d^7$ | $t_{2g}^6 e_g^1$ | $-18Dq + 3P$ |
| | | | | | |
| $d^8$ | $t_{2g}^6 e_g^2$ | $-12Dq + 3P$ | $d^8$ | $t_{2g}^6 e_g^2$ | $-12Dq + 3P$ |
| $d^9$ | $t_{2g}^6 e_g^3$ | $-6Dq + 4P$ | $d^9$ | $t_{2g}^6 e_g^3$ | $-6Dq + 4P$ |
| $d^{10}$ | $t_{2g}^6 e_g^4$ | $0Dq + 5P$ | $d^{10}$ | $t_{2g}^6 e_g^4$ | $0Dq + 5P$ |

## Tetrahedral Geometry

The CFT approach can be easily extended to other geometries and the next most important case is **tetrahedral**.

The $\Delta_{td}$ splitting of the d-orbitals is much smaller than $\Delta_o$ on the order of: $\Delta_{td} = 4/9\Delta_o$ since there are only **four** ligands around a tetrahedral complex compared to six on octahedral complexes, which reduces overall repulsion, and since the ligands are not as efficiently directed at the d-orbitals.

The scheme below depicts the Cartesian coordinate system with a hypothetical cube drawn around it, and then next a tetrahedral metal complex at the origin of the coordinate system, where the ligands are found at the corners of the hypothetical cube. Since the ligands are at the corners of the cube, no d-orbitals are directed towards them.

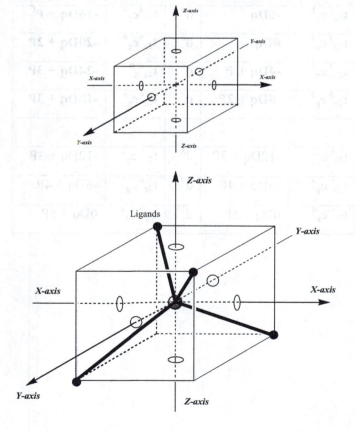

This can be illustrated by drawing a couple of representative d-orbitals into this scheme. For illustrative purposes and brevity, only the $dz^2$ and the $dxz$ orbitals are shown, representative of the e set and the t set of orbitals.

As can be discerned from the complicated scheme below, the e set of orbitals ($dz^2$ and the $dx^2-y^2$) have a lesser interaction with the ligands (1/2 face of cube away) than does the t set of orbitals ($dxy$, $dyz$, $dxz$) (1/2 of edge away).

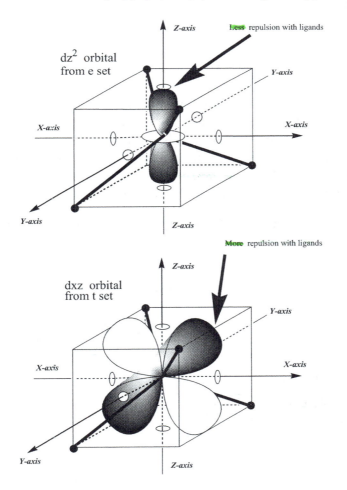

Since this is the reverse of the octahedral case, the d-orbital splitting diagram is reversed as seen on the next page.

Overall, the d-orbital splitting is not nearly as strong as in the octahedral case!

Since $\Delta_{td}$ is so small, *tetrahedral metal complexes are always considered to be weak field-high spin cases.*

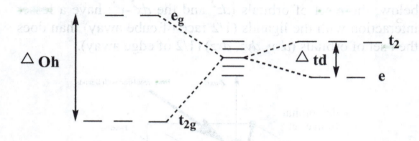

This can be illustrated with the tetrahedral complex ion [CoCl$_4$]$^{2-}$. The ion contains Co$^{2+}$. You should be able to determine that for yourself at this point. The Co(II) ion has a d$^7$ electron configuration. The splitting of the d-orbitals will be small since this is a tetrahedral complex, giving the weak field/ high spin case shown in the scheme below.

The compound [CoCl$_4$]$^{2-}$ exhibits ***paramagnetism***, with three unpaired electrons.

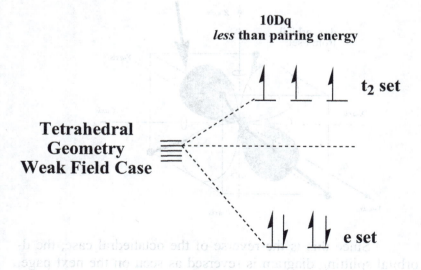

The **square planar case** can be considered as an extension of the octahedral field, where we remove the two ligands from the Z-axis.

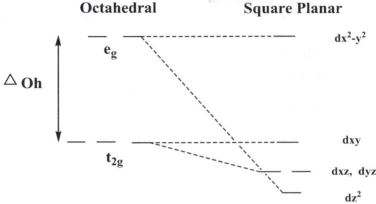

Consequently, repulsion of an electron in the $dz^2$ orbital will no longer be equivalent to that experienced by an electron in the $dx^2-y^2$ orbital, and the end result is shown above. In this text, the only square planar complexes will be for $d^8$ complexes, i.e. nearly all four coordinate complexes exhibit tetrahedral geometry except for $d^8$ metal complexes, which may be **tetrahedral or square planar**.

An example of a square planar complex is $[Ni(CN)_4]^{2-}$. The Ni(II) is $d^8$, so the splitting diagram is:

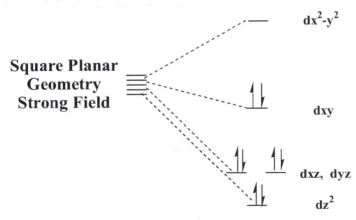

In this case, there are no unpaired electrons; the compound is considered to be ***diamagnetic***.

## The Jahn-Teller Effect

The ***Jahn-Teller theorem*** states that any species with an electronically ~~degenerate~~ ground state will distort to remove the degeneracy. Compounds exhibit the Jahn-Teller effect by displaying a distortion of their coordination geometry as the degeneracy of the ground state is "broken".

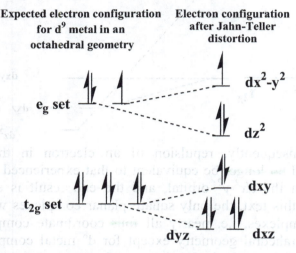

For transition metal complexes the Jahn-Teller effect is observed particularly for $d^9$ metal complexes in an ~~octahedral~~ field, and one of the best examples is the $[Cu(H_2O)_6]^{2+}$ ion. The bonds along the z-axis are ~~longer,~~ "~~lifting the degeneracy~~" of the $dz^2$ and the $dx^2-y^2$ orbitals.

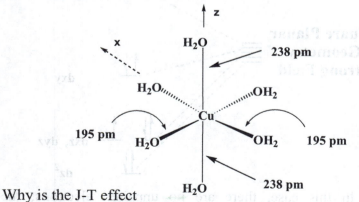

Why is the J-T effect also seen in some ~~high~~ spin $d^4$ and ~~low~~ spin $d^7$ cases?

## Trends in Crystal Field Theory

A trend in the value of $\Delta_o$ can be determined for metal ions which is independent of the ligands. This trend is:

Mn(II) < Ni(II) < Co(II) < Fe(III) < Cr(III) < Co(III) < Ru(III) < Mo(III) < Rh(III) < Ir(III) < Pt(IV). The trend follows roughly the oxidation state of the metals (+3 has a ~~higher~~ field than +2 for the same metal and ligand), then it increases down a group.

Generally, $\Delta_o$ increases with increasing oxidation state of the metals but, this is an empirical generalization and **crystal field theory just cannot explain the magnitudes of $\Delta_o$ values**. Generally, complexes of the $2^{nd}$ and $3^{rd}$ row of the periodic table almost always form **strong** field complexes, so Crystal Field Theory is best for $1^{st}$ row transition metal complexes.

## Magnetism

Compounds are considered to be ~~diamagnetic~~ when they have ~~no~~ unpaired electrons, but ~~paramagnetic~~ when they have ~~one or more~~ unpaired electrons. Since electrons have negative charge and are moving around an atom this constitutes an electric field, and electric currents set up magnetic fields.

An atom will exhibit a magnetic field when it is paramagnetic, but in a diamagnetic compound the paired electrons in an orbital cancel each other out so that no resultant magnetic field can be measured.

Paramagnetic compounds align these ~~unpaired~~ electrons in an applied external magnetic field, and this is how paramagnetism is measured. *~~Paramagnetic compounds are attracted to an applied external magnetic field.~~*

Diamagnetic compounds oppose the applied magnetic field (***Lenz's Law***) and are thus repelled by the applied external magnetic field.

The result is that ~~paramagnetic~~ compounds are ~~attracted~~ to an applied external magnetic field and become apparently ~~heavier~~ when weighed on a magnetic susceptibility balance, and ~~diamagnetic~~ compounds are ~~repelled~~ by an applied external magnetic field and become apparently ~~lighter~~ when weighed on a magnetic susceptibility balance.

Magnetic moments can then be calculated for each compound measured on a magnetic susceptibility balance. To a first approximation the magnetic moments can be calculated for a metal complex by using the number of unpaired electrons (n).

The following table shows some calculated and observed values for the magnetic moments of some metal complexes with their number of unpaired electrons.

**Table 5.3 Calculated and Observed Values of Magnetic Moments**

| Complex | n | $\mu_{Calculated}$ | $\mu_{Observed}$ |
|---|---|---|---|
| $[Cu(H_2O)_6]^{2+}$ | 1 | 1.73 | 1.75 |
| $[Ni(H_2O)_6]^{2+}$ | 2 | 2.83 | 2.83 |
| $[Co(H_2O)_6]^{2+}$ | 3 | 3.87 | 4.85 |
| $[Mn(H_2O)_6]^{3+}$ | 4 | 4.90 | 4.93 |
| $[Fe(H_2O)_6]^{3+}$ | 5 | 5.92 | 5.40 |

One method of determining the number of unpaired electrons is by looking at the magnetic properties of the compounds. A simple technique to determine a magnetic moment (the GOUY METHOD using a Gouy balance) involves weighing the sample in the presence and absence of a strong magnetic field. By careful calibration using a known standard, such as $HgCo(SCN)_4$ the number of unpaired electrons can be determined.

## Color

Colored compounds absorb visible light. The color perceived is the sum of the light that isn't absorbed by the complex. (This means that we see the color of a metal complex as the complimentary color of the absorbed species of light!) The amount of absorbed light versus wavelength is an absorption spectrum for a complex.

The color of a transition metal complex depends on the metal, its oxidation state, and its ligands. The pale blue transition metal species $[Cu(H_2O)_4]^{2+}$ can be converted into the dark blue $[Cu(NH_3)_4]^{2+}$ by adding $NH_3(aq)$. A partially filled set of d orbitals is usually required for a complex to be colored. *So, $d^0$ and $d^{10}$ metal ions are usually colorless.* Exceptions to this are $MnO_4^-$, which is Mn(VII), and $CrO_4^{2-}$, which is Cr(VI). These colors are from ***charge-transfer***.

For example, the absorption spectrum for $[Ti(H_2O)_6]^{3+}$ has a $\lambda_{max}$ (maximum absorption) at 510 nm, which is in the green-yellow part of the visible spectrum, so, the complex transmits all light except green-yellow. Therefore, we see the complex as the complimentary color of this yellow-green absorption, which is purple. See the color wheel on the front cover of the book. (The absorption peaks are often very broad so the colors expected from comparison with the color wheel may not match exactly.)

*For some various transition metal species:*

| ~Color absorbed | ~Wavelength absorbed | ~Color observed |
|---|---|---|
| Violet | 410nm | yellow |
| Blue-green | 460nm | orange |
| Green | 500nm | red |
| Red | 660nm | blue |

Since $E=h\nu$ and $E=hc/\lambda$, then
the shorter the wavelength ($\lambda$) absorbed, the larger is $\Delta_o$.

## Kinetics

The rate of ligand exchange can be fast for a metal complex (***kinetically labile***) or it can be very slow (***kinetically inert***). The rate of ligand exchange can depend on many factors such as the exchange mechanism, type of ligand, solvent, temperature, etc. An important factor is the crystal field stabilization energy of a metal complex.

If the metal is highly stabilized, such as $Co^{3+}$, which has $t_{2g}^6$ electron configuration (low spin) and 24Dq CFSE, or $Cr^{3+}$ which has $t_{2g}^3$ electron configuration (low spin) and 12Dq CFSE, then it is kinetically inert. Complexes with other electron configurations such as $d^4$ and $d^9$ are very kinetically labile. Why?

Inert = $d^3$, $d^6$, $d^8$ (low spin)   Labile = $d^4$, $d^9$

## Limitations of the Crystal Field Theory

The Crystal Field Theory is a metal-based theory, and does not delve into the effects of various ligands on the crystal field strength. Generally, $\Delta_o$ increases with increasing oxidation state of the metals, however, this is an empirical generalization and crystal field theory just cannot explain the magnitudes of $\Delta_o$ values. Real progress in understanding can only be made by considering the bonding between ligands and metals.

There are other major problems with the crystal field theory. Ligands are not point charges, but crystal field theory treats ligands as point charges or dipoles. Even charged ligands such as the chloride ion have a polarized charge distribution when bonded to transition metal ions, so they can't be considered to be point charges either.

Crystal field theory does not take into account the overlapping of ligand orbitals with metal-based orbitals in a more covalent fashion, such as in the bonding of neutral molecules such as carbon monoxide with transition metals.

Another problem is the lack of distinction between ~~strong~~ field ligands and ~~weak~~ field ligands. Why does a ligand such as chloride and water form weak field complexes, yet ligands such as the organophosphines and carbon monoxide form strong field complexes?

The crystal field theory has no explanation for these observations, and as a matter of fact, it fails completely. *How can the anionic $Cl^-$ ion function as a weak field ligand when a neutral covalent molecule such as carbon monoxide acts as a strong field ligand?*

Clearly the crystal field theory is incomplete and either needs to be modified or discarded. Since it is so useful, it has been modified and has evolved instead into ***Ligand Field Theory.***

**Ligand Field Theory**

Van Vleck adopted Ligand Field Theory in 1935 as a modification of Crystal Field Theory in order to account for more covalency in bonding of ligands to metal, which is lacking in the purely electrostatic Crystal Field Theory. *It has been established that the ~~ability of ligands to cause a large splitting of the energy between the orbitals is essentially independent of the metal ion~~.*

The ***Spectrochemical Series*** is a list of ligands ranked in order of their ability to cause large orbital separations. A shortened list includes:

$I^-$ < $Br^-$ < $\underline{S}CN^-$ ~ $Cl^-$ < $F^-$ < $OH^-$ ~ $\underline{O}NO^-$ < $C_2O_4^{2-}$ < $H_2O$ < $\underline{N}CS^-$ < $EDTA^{4-}$ < $NH_3$ ~ pyr ~ en < bipy < phen << $PR_3$ < $CN^-$ ≈ $NO$ ≈ $CO$

Take, for example, two different metal complexes of $Fe^{2+}$ - $d^6$, which are: $[Fe(H_2O)_6]^{2+}$ and $[Fe(CN)_6]^{4-}$.

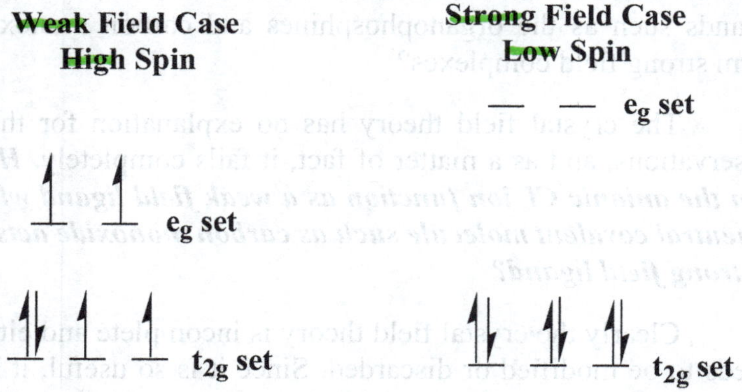

**Weak Field Case**      **Strong Field Case**
**High Spin**      **Low Spin**

⎯ ⎯    $e_g$ set

↑ ↑    $e_g$ set

↑↓ ↑ ↑    $t_{2g}$ set      ↑↓ ↑↓ ↑↓    $t_{2g}$ set

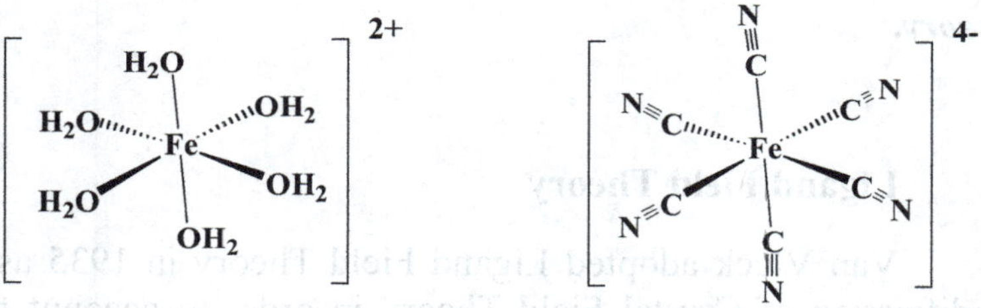

$$\left[\begin{array}{c} H_2O \\ H_2O\cdots Fe\cdots OH_2 \\ H_2O \quad OH_2 \\ OH_2 \end{array}\right]^{2+} \qquad \left[\begin{array}{c} N\equiv C \\ N\equiv C\cdots Fe\cdots C\equiv N \\ N\equiv C \quad C\equiv N \\ C\equiv N \end{array}\right]^{4-}$$

The diagram on the left represents the case for the aqua ion (small Δ) and on the right that of the hexacyano ion (large Δ).

What this means is that if one uses a technique, such as magnetic susceptibility, that can detect the presence of unpaired electrons in each compound, then one will find that $[Fe(H_2O)_6]^{2+}$ has four unpaired electrons, while in the latter compound $[Fe(CN)_6]^{4-}$ there will be no unpaired electrons.

This accounts for the terms *high spin* (most unpaired electrons and most paramagnetic) and *low spin* (least unpaired electrons).

The terms weak field and strong field give an indication of the splitting abilities of the ligand.

Water always gives rise to ~~small splittings~~ of the energy levels of the d orbitals for first row transition metal ions and hence is referred to as a ~~weak~~ field ligand. Conversely, $CN^-$ is a strong field ligand, since it causes large splittings of the energy levels of the d-orbitals.

How do you know if a ligand is strong or weak field without the use of the spectrochemical series in hand? **The general rule of thumb is that CO, $CN^-$, phosphines and $NO^+$ are all strong field ligands.** The halides are weak field ligands, and the nitrogen-based ligands are probably weak field but can be ~~strong field ligands~~ in certain cases. We will discuss this further in Chapter 6, and explain the spectrochemical series using molecular orbital theory.

As has been mentioned by other inorganic chemists, the valence-bond theory picture and the ligand field theory picture can both be considered to be specializations of the molecular orbital method.

Some aspects of the material in the next chapter on MO theory are within the boundaries of Ligand Field Theory. We have decided to combine the material and present the quantum mechanical aspects of Ligand Field Theory and Molecular Orbital Theory at one time in the same chapter.

# Inorganic Chemistry: Introduction to Coordination Chemistry

## Terms and Definitions for Chapter 5:

***Orbital Hybridization Concept***
***Hybrid orbitals***

***Electronic Configuration***

***Crystal Field Theory***

***Paramagnetism***

***Visible Absorption Spectra***
***Charge Transfer***
***Simple Point Charges***
***Electrostatic Potential***
***Cartesian Coordinate System***

***Crystal Field Splitting***
***10dq***
***$e_g$ Set***
***$t_{2g}$ Set***
***Crystal Field Stabilization Energy* (CFSE)**
***Weak Field Case***
***Strong Field Case***
***Hund's Rule***
***High Spin***
***Low Spin***

***Kinetically Labile***
***Kinetically Inert***

***Diamagnetic***
***Paramagnetic***

***Ligand Field Theory***
***Spectrochemical Series***

***Jahn-Teller Theorem***

# Inorganic Chemistry: Introduction to Coordination Chemistry

## Chapter 5 Problems and Exercises

1. What is the color of an aqueous solution containing $Zn(NH_3)_4^{2+}$? (a) blue (b) yellow (c) red (d) colorless

2. Which pair has a $d^{10}$ electron configuration?
(a) $Cu^{2+}$ and $Ni^{2+}$ (b) Co and Ni
(c) Zn and $Cu^{2+}$ (d) Cu and Zn (e) Cu+ and $Zn^{2+}$

3. What are the electron configurations of $Fe^{2+}$ and $Co^{3+}$, respectively?

(a) both $d^7$ (b) $d^7$ and $d^8$ (c) both $d^6$

(d) $d^6$ and $d^8$ (e) both $d^5$

4. Which of the following complexes contains a $d^3$ metal ion? Give the number of d-electrons for each metal.
    (a) $K_2[Ni(CN)_4]$
    (b) $[Ni(NH_3)_6]Cl_2$
    (c) $K_4[Fe(CN)_6]$
    (d) $[Co(OH_2)_6](ClO_4)_2$
    (e) $[Cr(OH_2)_5Cl]Cl_2$

5. Predict the number of unpaired electrons in $[Cr(CN)_6]^{4-}$ and $[Cr(OH_2)_6]^{2+}$, respectively. Draw the d-orbital splitting diagrams.

6. Predict the number of unpaired electrons in the tetrahedral complex ion $[MnCl_4]^{2-}$. Draw the d-orbital splitting diagram.

7. Rank the following complex ions from the lowest d-orbital splitting energy to the highest.
(a) $[Co(OH_2)_6]^{3+}$ (b) $[Co(ox)_3]^{3-}$ (c) $[CoCl_6]^{3-}$
(d) $[Co(NH_3)_6]^{3+}$ (e) $[Co(CN)_6]^{3-}$

8. Rank the following complex ions from the lowest d-orbital splitting energy to the highest.
(a) $[Fe(NH_3)_6]^{2+}$ (b) $[Fe(CN)_6]^{4-}$ (c) $[Fe(OH_2)_6]^{2+}$
(d) $[FeCl_6]^{4-}$ (e) $[FeCl_4]^{2-}$

# Inorganic Chemistry: Introduction to Coordination Chemistry

9. For which one of the following would it not be possible to distinguish between high-spin and low-spin complexes in octahedral geometry?
(a) Ni(II)  (b) Co(III)  (c) Fe(II)  (d) Co(II)  (e) Cr(II)

10. The complex ion $[Cr(OH_2)_6]^{2+}$ has four unpaired electrons. Which statements are true?
(a) the complex is high-spin.
(b) the complex is low-spin.
(c) the complex is diamagnetic.
(d) $\Delta_o$ is very large.
(e) the water ligands are difficult to remove.

11. Predict the total number of d-electrons in a complex having one unpaired electron in a strong octahedral field and three unpaired electrons in a weak octahedral field.

12. An octahedral complex is high-spin when the value of $\Delta_o$ is _____ than the energy required to _____ the electrons. (hint-Hund's rule)

13. Determine which of these metal complexes are high spin.
$[CoF_6]^{3-}$      $\mu_{Obs} = 5.3$
$[Fe(CN)_6]^{3-}$   $\mu_{Obs} = 2.3$
$[Co(NO_2)_6]^{4-}$ $\mu_{Obs} = 1.8$

14. Which complex ion should absorb visible light of the shortest wavelength, $[Cr(H_2O)_6]^{3+}$, or $[CrCl_6]^{3-}$?

15. For $Co^{3+}$, the pairing energy is 252 kJ/mol and the d-orbital splitting energies produced by F- ligands is 155 kJ/mol, and the d-orbital splitting energies produced by $NH_3$ ligands is 276 kJ/mol. Sketch the d-orbital splitting diagrams for the two complexes $[CoF_6]^{3-}$ and $[Co(NH_3)_6]^{3+}$, and place electrons in the correct d-orbitals. Which are high spin or low spin complexes?

# Inorganic Chemistry: Introduction to Coordination Chemistry

16. On the basis of the given colors for the following complexes, list the ligands $NH_3$, $H_2O$, $Cl^-$, and $NCS^-$ in order of increasing field strength, and then compare your list with the spectrochemical series.

$[Co(NH_3)_6]^{3+}$      orange-yellow
$[Co(NH_3)_5Cl]^{2+}$      purple
$[Co(NH_3)_5(NCS)]^{2+}$      orange
$[Co(NH_3)_5(H_2O)]^{3+}$      red

17. Rank the following complex ions in order of increasing $\Delta_o$ and the energy of visible light absorbed: $[Cr(NH_3)_6]^{3+}$, $[Cr(H_2O)_6]^{3+}$, $[Cr(Br)_6]^{3-}$.

18. Rank the following complex ions in order of decreasing $\Delta_o$ and the energy of visible light absorbed: $[Cr(en)_3]^{3+}$, $[Cr(CN)_6]^{3-}$, $[Cr(Cl)_6]^{3-}$.

19. The complex $[Cr(H_2O)_6]^{3+}$ is violet. A second $CrL_6$ complex is green. The ligand is probably... (a) $CN^-$ (b) $Cl^-$

20. Draw a d-orbital diagram for the $Mn^{3+}$ ion (high spin) in an octahedral environment. Would you suspect that Jahn-Teller distortion could occur?

21. Draw a d-orbital diagram for the $Ni^{3+}$ ion (low spin) in an octahedral environment. Would you suspect that Jahn-Teller distortion could occur?

22. Explain why coordination compounds containing $Cu^{2+}$ are colored but coordination compounds containing $Cu^+$ are not.

23. Complexes with the electron configurations $d^4$ and $d^9$ are usually very kinetically *labile*. Why?

24. Complexes with the electron configurations $d^3$ and $d^6$ are usually very kinetically *inert*. Why?

# Inorganic Chemistry: Introduction to Coordination Chemistry

**Practice Quiz**

1. What is the oxidation state (use roman numerals please) of the metal and the number of d electrons in:

$[CoCl_6]^{3-}$ = _____ and it has _____ d electrons
$[Cr(CN)_6]^{4-}$ = _____ and it has _____ d electrons
$[CuCl_4]^{2-}$ = _____ and it has _____ d electrons
$[Cr(OH_2)_6]^{2+}$ = _____ and it has _____ d electrons

2. Draw the five d-orbitals with respect to their alignment with the x, y, and z-axes.

Label the $e_g$ and the $t_{2g}$ set.

3. Predict the number of unpaired "d" electrons in $[W(CN)_6]^{4-}$ and $[W(OH_2)_6]^{2+}$, respectively. Draw the d-orbital splitting diagram for these octahedral complexes.

_____ 4. Predict the number of unpaired electrons in the tetrahedral complex ion $[CuCl_4]^{2-}$.
(A) 1  (B) 2  (C) 0  (D) 3  (E) 4

_____ 5. Which of the following complex ions has the smallest d-orbital splitting energy ($\Delta_o$) ?
(A) $[Rh(OH_2)_6]^{3+}$  (B) $[Rh(ox)_3]^{3-}$  (C) $[RhCl_6]^{3-}$
(D) $[Rh(NH_3)_6]^{3+}$  (E) $[Rh(CN)_6]^{3-}$

_____ 6. Which of the following complex ions has the highest d-orbital splitting energy ($\Delta_o$) ?
(A) $[Os(NH_3)_6]^{2+}$  (B) $[Os(CN)_6]^{4-}$  (C) $[Os(OH_2)_6]^{2+}$
(D) $[OsCl_6]^{4-}$

7. For what numbers of d-electrons would it not be possible to distinguish between high-spin and low-spin complexes in octahedral geometry?

## Inorganic Chemistry: Introduction to Coordination Chemistry

8. The complex ion $[Mo(OH_2)_6]^{2+}$ has four unpaired electrons. This means that (circle all that are correct)....
   (A) the complex is high-spin.
   (B) the complex is low-spin.
   (C) the complex is diamagnetic.
   (D) the complex is paramagnetic
   (E) $\Delta_o$ is larger than the pairing energy
   (F) $\Delta_o$ is smaller than the pairing energy.

_____9. Predict the total number of d-electrons in a complex having two unpaired electrons in a strong octahedral field and four unpaired electrons in a weak octahedral field.
(A) 7  (B) 5  (C) 6  (D) 8  (E) 4

_____10. An octahedral complex is high-spin, weak field when the value of $\Delta_o$ is larger than the energy required to pair up the electrons. (TRUE or FALSE)

11. How many unpaired electrons would $[Tc(CN)_6]^{3-}$ and $[TcCl_6]^{2-}$ have?

Give electronic configurations for each ion in terms of the $t_{2g}$ and $e_g$ sets. Carefully rationalize your answers by drawing the d-orbital splitting diagrams for each one and tell if it is high field (large $\Delta o$) or weak field (small $\Delta o$).

BONUS question: $Na_2[Pt(CN)_4] \cdot 3H_2O$ is diamagnetic.

Speculate on the basic geometry of the complex anion, $[Pt(CN)_4]^{2-}$, by using a d-orbital crystal field splitting diagram. Briefly explain your answer. Is it tetrahedral or square planar?

# Chapter 6. Molecular Orbital Theory

The molecular orbital approach spreads electrons out over many atoms (like resonance) and thus has the effect of giving a ~~more stable~~ bonding situation for molecules, and even transition metal complexes.

To be worthwhile, the molecular orbital approach has to explain bonding for weak and strong field ligands, and also not lose the usefulness of crystal field theory. Some of the strongest aspects of the molecular orbital theory are that it can explain the ~~bonding situation~~ for metal carbonyls and strong field ligands and explain the overall spectrochemical series.

In MO theory, we will start with looking at ~~combinations~~ of atomic orbitals on the ligands and the metal center and see what kind of orbital overlaps are allowed. This is only an introductory level college course textbook, so some readers will be unfamiliar with a lot of molecular orbital theory. That is okay, this is presented and based on material about atomic orbitals (s, p, d) that the reader should have been exposed to in a freshman chemistry course, and then we'll go from there.

In ~~Ligand Field Theory~~, along with basic Molecular Orbital Theory, we start our systematic discussion of bonding by constructing *linear combinations of atomic orbitals* (LCAO's) based on atomic orbitals from each of the bonding atoms.

The best way to present this material is to first look at the simplest systems, such as the simple ~~diatomic~~ molecules. This allows us to systematically build more complex sets of LCAO's, and help us to distinguish between sigma and pi interactions between these linear combinations of atomic orbitals.

The difference between core and valence electrons can then also be distinguished in this manner, and we can then direct our attention on the outermost LCAO's, which are often called the "frontier orbitals".

## σ (sigma) Bonding

Electron density lies between the nuclei in molecular compounds for any sigma bond. A sigma bond may be formed by linear combinations of **s-s, p-p, s-p, s-d, p-d, d-d,** or hybridized orbital overlaps, and shortly we will work through examples of the first two of these (s-s, and p-p).

The bonding molecular orbitals produced by the linear combination of atomic orbitals approach are lower in energy than the original atomic orbitals.

The antibonding MO's are higher in energy than the original atomic orbitals.

## Bond Order

Bond order is the measure of the stability (or reactivity) of a chemical covalent bond, and electrons in bonding orbitals stabilize the bond, whereas electrons in antibonding orbitals destabilize the bond.

The higher the bond order, the more stable the bond; and the bond order can be calculated by the following equation:

$$BO = \frac{(e^- \text{ in bonding orbitals}) - (e^- \text{ in antibonding orbitals})}{2}$$

**Simple Inorganic Molecules (diatomic molecules)** For a simple diatomic molecule such as $H_2$, the molecular orbital description is not all that complicated.

For $H_2$, the covalent bond is formed by simple overlap of s-orbitals to form a **sigma** (σ) bond. The sigma bond essentially allows the electron pair to travel over a larger area, and thus it has less energy than the simple atomic s-orbitals it is derived from. The scheme below shows the formation of a sigma bond (constructive overlap), and the concomitant antibonding orbital that also forms (destructive overlap).

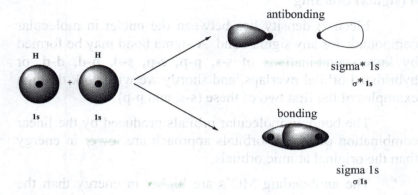

The formation of He₂ is not observed in nature and our theory shows that the antibonding orbital would be filled, and therefore there would be no net bonding interaction.

The molecular orbital energy diagrams for H₂, and He₂ are shown in the following scheme.

Other diatomic inorganic molecules that are commonly found in the environment, or can be readily synthesized are N₂, O₂, and F₂. We drew the Lewis electron dot structures of these molecules earlier on, and we should remember that N₂ has a triple bond when the Lewis structure is drawn.

What would molecular orbital theory tell us? Each nitrogen atom has seven total electrons, but two are used to fill the 1s core, and therefore nitrogen (Group V) has five valence electrons in 2s and 2p orbitals.

We discuss this next.

The three p orbitals, when all are shown in a three-dimensional fashion around the central atom, take up much of the space, as can be seen in the scheme below. The three p orbitals, by convention, lie along the x, y, and z axes of the Cartesian coordinate system.

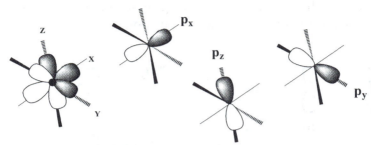

The direct overlap of orbitals between the bonding atoms in molecular orbital theory results in a σ-bond (sigma) regardless of the type orbitals that are used to construct the molecular orbital.

Thus, a σ-bond can be formed by the **constructive** overlap of two in-phase p orbitals on bonding atoms in a diatomic molecule.

This kind of **sigma overlap** also results in **destructive** interference by out-of-phase orbitals to produce a σ* antibonding molecular orbital. These molecular orbitals are shown in the following scheme:

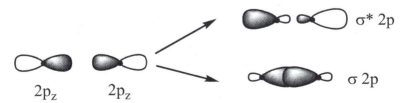

The remaining two p orbitals from one atom in a diatomic molecule cannot directly overlap with the other remaining p orbitals on the second atom to form a sigma bond, but they can overlap in a quite different fashion.

## π (pi) Bonding

This side-to-side overlap is known as a pi (π) molecular orbital. The two possible p-π molecular orbitals for a diatomic molecule are ~~degenerate~~ in energy, i.e., they have the ~~same~~ energy. One of the two degenerate p-π molecular orbitals is shown in the following scheme, as well as the resulting antibonding p-π* molecular orbital.

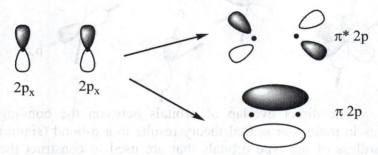

The resulting antibonding p-π* molecular orbitals are consequently of ~~higher~~ energy than the bonding p-π molecular orbitals. The molecular orbital energy diagram for the p-orbital interactions to form molecular orbitals can now be shown in the following scheme.

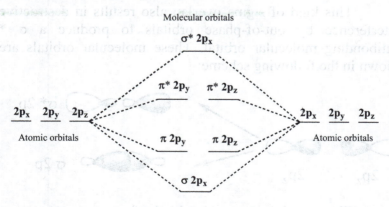

If we now consider a real molecule (let's take the case of $F_2$) then we must add in the electrons into the molecular orbitals.

Let us now consider all of the electrons for $F_2$, the core electrons as well as the valence electrons.

The molecular orbital energy diagram shown below illustrates all of the points we have discussed so far.

The bond order is 1. This can be calculated using the bond order equation shown earlier. The molecule has 10 electrons in bonding orbitals and 8 electrons in antibonding orbitals.

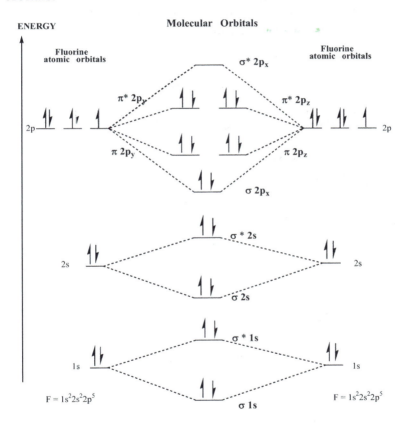

The electrons in LCAO's produced by the 1s orbitals are core electrons. The electrons in the LCAO's produced by the 2p orbitals are in *"frontier orbitals"* which are the **highest occupied molecular orbitals (HOMO)**, and the **lowest unoccupied molecular orbitals (LUMO)** are unfilled.

Molecular oxygen ($O_2$) has some interesting properties. The Lewis structure shows the presence of a double bond. Draw it. ::O=O::

Even though the Lewis structure shows the presence of a double bond, and no unpaired electrons, the physical data contradicts this drawing. When placed in a magnetic field, liquid oxygen is attracted to the poles of a magnet. **This shows that $O_2$ is paramagnetic, and has unpaired electrons!!** Molecular orbital theory on the other hand predicts that it should have ~~unpaired~~ electrons!!

Thus for molecular $O_2$:

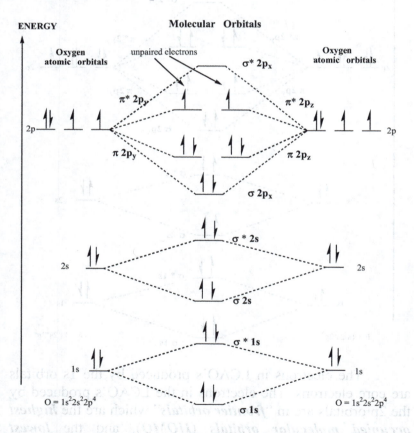

What is the bond order for molecular $O_2$ using MO theory?

## Molecular Orbitals for d-block elements

The transition metal center has s, p, and d orbitals that are available to provide orbital overlap with ligand-based orbitals. The ligands themselves may have "hybrid orbitals" for the lone pairs of electrons on the ligating atoms, such as $sp^3$, $sp^2$, or sp hybrid orbitals, and these can be used to overlap the atomic orbitals on the metal. Orbitals are drawn with different shadings on the different lobes of p and d orbitals, these are called angular overlap shadings, and may be alternatively designated with a + or − sign.

Orbitals can only overlap and "merge" together if they have the same overlap symbol (said to be "in-phase"), otherwise they are considered to be "out of phase". This is analogous to constructive and destructive interference of waves. So, let us look at that in phase/out of phase situation for different orbitals. As shown in the diagram below there is a net overlap for a d orbital with an s orbital, as is the case between d and p orbitals if they are correctly aligned.

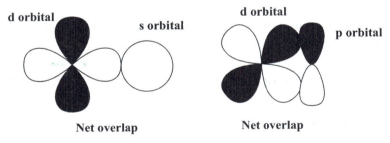

On the other hand, improper alignment of orbitals will lead to situation where the orbitals cannot merge, and there is no net overlap. This is seen in the scheme below:

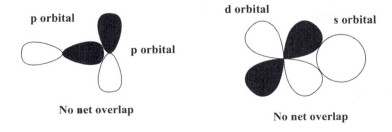

Now let us imagine a more complicated situation. Let us imagine a hypothetical octahedral metal complex with six ligands bound to the metal. MO theory says that we should be able to draw six molecular orbitals that would describe the six bonds. However, unlike valence bond theory, the bonds are not all of the same energy, and the bonds may be spread out over many atoms. Our bonding combination of orbitals is based on symmetry considerations. Thus:

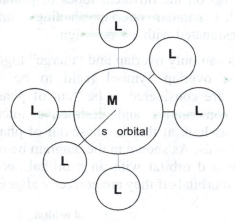

Here the metal s orbital (4s) can merge with six ligand lone pair orbitals (hybrid orbitals) to form a molecular orbital ($a_{1g}$) bonding combination. This is the lowest energy molecular orbital for bonding between ligands and the metal.

Next, we can construct a set of three molecular orbitals using three p orbitals on the metal, and the ligand orbitals:

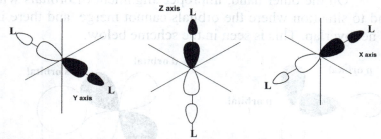

These three molecular orbitals are given the designation $t_{1u}$.

Lastly, the five d-orbitals form a set of two **bonding** molecular orbitals (**e_g set** with the $dz^2$ and the $dx^2-y^2$), and a set of three non-bonding orbitals (**t_{2g} set** with the **dxy, dxz, and the dyz orbitals**).

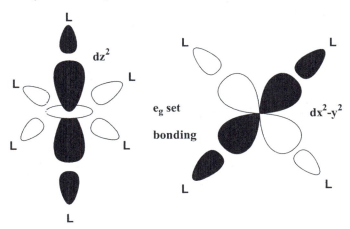

As you can see, there are six bonding molecular orbitals that can be formed from the metal orbitals; there are two bonding molecular orbitals and three non-bonding molecular orbitals.

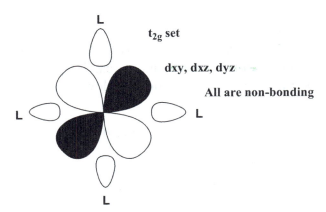

In MO theory for every bonding molecular orbital that can be constructed, there is a corresponding antibonding orbital of higher energy. The molecular orbitals from lowest energy to highest energy would be bonding, non-bonding, and then antibonding.

## σ-base ligands (σ-donor ligands) [NH₃]

The following is an energy level diagram for a ML$_6$ complex showing the sigma-bonding contributions.

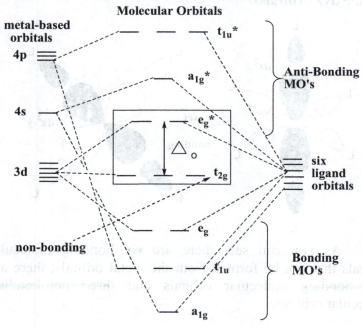

From the MO diagram above, one can see that even though a complicated MO diagram has evolved, the essential crystal field component remains intact, and $\Delta_o$ can be seen. The insert showing the frontier orbitals is shown below:

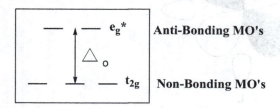

This MO model can be investigated even further for different types of ligands and for different complex geometries, but the main point has been established. MO theory can be developed and yet maintain the useful aspects of crystal field theory. (see problem 5)

## π-base ligands (π-donor ligands/σ-donor) [Cl]⁻

The lone pairs on a ligand such as the chloride ion essentially act as π-donors and contribute electron density onto the $t_{2g}$ orbitals based on the metal. These $t_{2g}$ orbitals become bonding orbitals along with the corresponding formation of the antibonding $t_{2g}^*$ orbitals at higher energy.

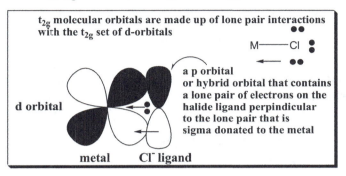

The effect of π-donor ligands such as chloride ion is to decrease $\Delta o$. The following is an energy level diagram for a $ML_6$ complex showing the sigma-bonding contributions, and the π-bonding contributions. (see problem 6)

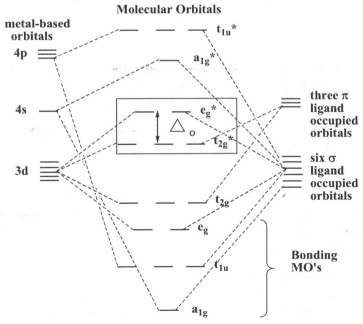

## $\pi$-acid ligands ($\pi$- acceptor ligands/$\sigma$-donor) [CO]

The following is an energy level diagram for a $ML_6$ complex showing the sigma-bonding contributions, and importantly, the $\pi$-bonding contributions for a $\pi$ - acceptor ligand. The empty $\pi$* orbitals on a ligand such as carbon monoxide act essentially as $\pi$-acceptors and siphon off electron density from the $t_{2g}$ orbitals based on the metal.

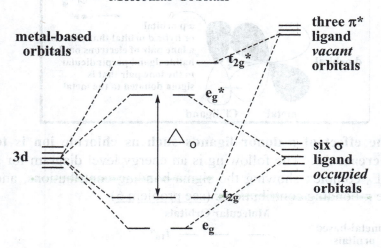

**Molecular Orbitals**

The effect of $\pi$-acceptor ligands such as carbon monoxide is to increase $\Delta o$. (see problem 7)

The next important point is how to describe the bonding for strong field ligands. Strong field ligands were actually discussed in Chapter 3 as *nonclassical ligands, $\pi$-bonding or $\pi$-acid -ligands*.

In Lewis acid/base theory, one can recall that a Lewis acid is an electron acceptor. *Nonclassical ligands are $\pi$-acid -ligands, and accept electron density from the metal by using appropriate $\pi$ orbitals*.

Carbon monoxide (CO) is the prototype $\pi$-acceptor ligand. Carbon monoxide has a triple bond between the carbon and oxygen atoms, and these bonding molecular orbitals have corresponding *empty* anti-bonding π* molecular orbitals that can be utilized in bond formation with transition metals that have partially filled d-orbitals.

One of these can be drawn in the manner shown below:

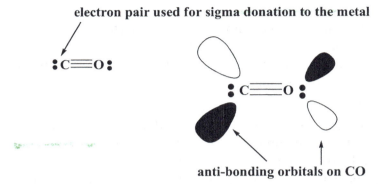

Carbon monoxide has three bonding orbitals that focus electron density mainly right between the carbon and oxygen atoms. The two anti-bonding orbitals (only one of which is shown in the diagram above) are empty, but have interesting overlap integrals that are strikingly like those of the d-orbitals.

If transition metals in the metal complexes have partially filled d-orbitals, then CO antibonding orbitals have the correct symmetry to merge with them. Carbon monoxide still possesses a lone pair of electrons on the carbon atom that can be donated to the metal. The π-bonding situation shown here is referred to as *backbonding*.

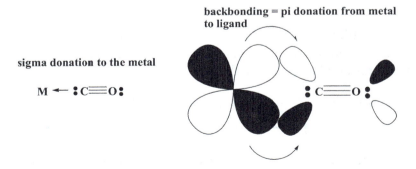

# Inorganic Chemistry: Introduction to Coordination Chemistry

These pi bonds make up the $t_{2g}$ set of molecular orbitals discussed earlier for the π-acid ligands such as CO.

Interestingly, this backbonding gives rise to two different bonding modes for CO to a transition metal.

The sigma donation is actually very weak since the lone pair on the carbon atom is held tightly (the lone pair on the oxygen atom is held even tighter---if you don't believe this calculate the Formal Charge on the oxygen atom) and thus the metal-carbon bond should be weak.

However, as we will see in the next chapter on organometallics, the metal-carbon bond is fairly strong, and metal carbonyl compounds are stable. The backbonding between the metal and the CO ligand, where the metal donates electron density to the CO ligand forms a *dynamic synergism* between the metal and ligand, which gives unusual stability to these compounds. This situation is shown below.

**Metal -CO bonding is a Dynamic Synergism**

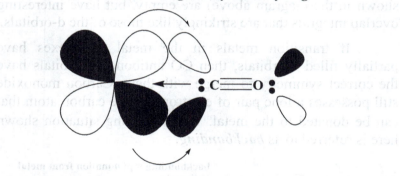

Utilizing Lewis electron dot structures and using a valence-bond approach (resonance forms) one could draw the structures and bonding like this:

A simple look at metal-carbonyl backbonding

M—C≡O:   ⟷   M=C=O:

This is actually instructive because it suggests that the carbon-oxygen <u>bond order decreases</u> from 3 to 2 <u>as backbonding increases</u>. This can be followed easily using infrared spectroscopy since carbonyl (C=O) groups strongly absorb infrared radiation.

(stretching frequencies-see the discussion on **pages 196-197**) Backbonding will be discussed further in Chapter 7.

The ~~unoccupied~~ carbon monoxide π-antibonding orbitals combine with the non-bonding $t_{2g}$ orbitals (that are filled or partially filled with electrons) and become bonding. ***This has the effect of lowering the $t_{2g}$ orbital energies, and subsequently increasing $\Delta o$.*** Thus, these type ligands (CO, $CN^-$ and $NO^+$) promote a ~~strong field~~.

## Organophosphine Ligands

Phosphine ligands are strong field ligands also, but they exhibit backbonding in a quite different fashion than carbon monoxide. They achieve backbonding by using either their empty 3d orbitals or their ~~empty sigma* orbitals~~. This issue has not been decided definitively and controversy exits as to which type of bonding shown here is actually the predominate one.

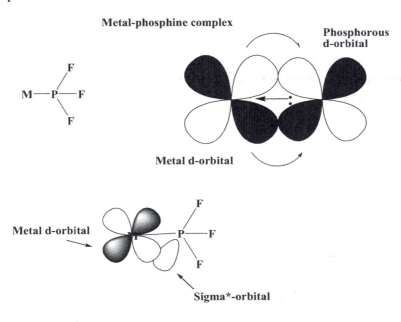

Interestingly, phosphine ligands can be modified so that they exhibit a wide range of properties. Two factors are most important: (1) the ~~electronegativity~~ (electron-withdrawing ability) of the substituents (the R groups), and also (2) their ~~steric bulk~~. As the electronegativity of the R groups increases on the phosphine, the ~~π-acceptor~~ properties of the phosphine ligand increases.

Steric bulk can be measured by the *"cone angle"* of the phosphine ligand. Often, ~~larger and bulkier~~ phosphine ligands can promote active sites on the metal, which may increase catalytic activity. This is often seen with rhodium catalysts such as the ***Wilkinson's hydrogenation catalyst***.

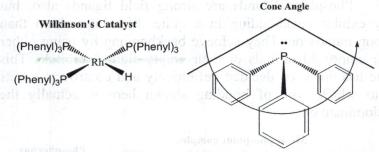

The organophosphine ligands can be synthesized so as to achieve a chiral phosphorous center which influences the ~~hydrogenation reaction~~. This was the basis for one synthesis of L-Dopa using the chiral DIOP ligand.

## Alkene Ligands

In 1827 Zeise prepared a platinum compound that he formulated as KCl· PtCl$_2$ · EtOH, but its structure was not established until the 1950's when it was shown to be a metal-alkene complex instead. The structure of the compound known as *Zeise's Salt*, is shown below.

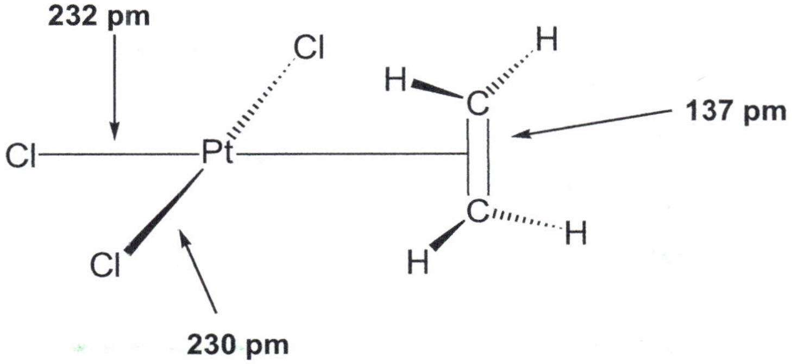

Some of the important features of this compound are that the alkene binds through both carbon atoms instead of only one atom, like most ligands, and that the alkene has a lowered carbon-carbon bond distance that suggests that the bond order is decreased from a double bond to somewhere intermediate between a double and single carbon-carbon bond.

The normal bond distance between the carbon atoms in a double bond is typically about 134 pm, and the bond distance between the carbon atom in Zeise's Salt is 137 pm. The typical carbon single bond distance is around 154 pm.

A similar compound is the amine complex trans-PtCl$_2$(NMe$_2$H)(C$_2$H$_4$) which has a C-C bond distance of 147 pm, which is very close to that of a C-C single bond.

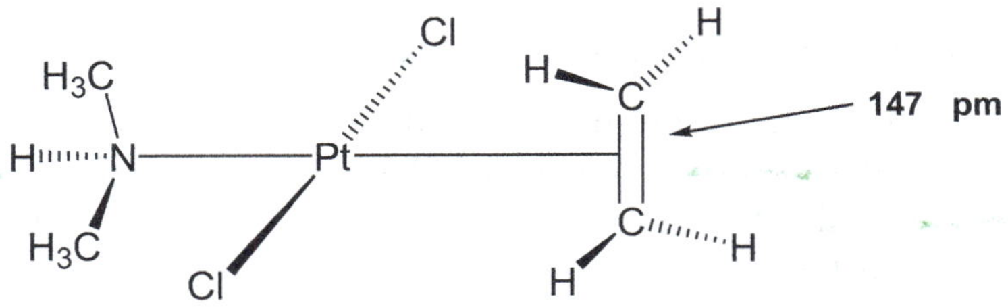

The bonding description is also very interesting, and the features of this bonding are shown below. Importantly, MO theory provides us with a framework in which to describe the backbonding in metal alkene complexes.

**Sigma Bond Component**

Filled alkene π bond

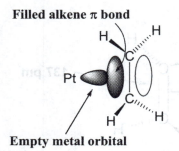

Empty metal orbital

**Pi Acceptor Bond Component**

Empty alkene π* orbital

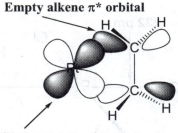

Filled metal d-orbital

The π-bond in the alkene acts as a sigma donor (like a lone pair) to form the sigma bond to the metal, and then the metal can participate in backbonding by pushing electron density in a d-orbital back onto the π* antibonding orbital of the alkene, which reduces the bond order of the alkene.

Thus, the alkene acts as a σ-donor and a π-acceptor ligand.

Alkene-metal complexes are important for many commercial applications such as in the production of high-density polypropylene using *Ziegler-Natta* **catalysts**.

## Summary: the Spectrochemical Series

After discussing the Ligand Field/Molecular Orbital treatment of the covalent bonding of ligands to transition metals, some useful observations emerge, and we are now in a position to explain the spectrochemical series.

The order of the ligands in the spectrochemical series is partly the strength with which they can participate in σ-bonding with the metal in question. More importantly however, is the participation and type of π-bonding that occurs between the ligand and the metal, for that is what determines the value of $\Delta o$.

When compared, using the common $e_g^*$ MO as our standard, then we see the differences in magnitude of $\Delta o$ that the different ligands infer on metals in an octahedral ligand field. This comparison is shown below using only the MO's based on the d-orbitals.

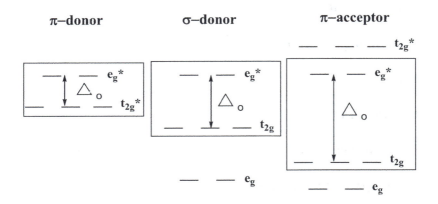

The overall order of the spectrochemical series may be rightly interpreted as being dominated by π effects.

The general series can then be seen as follows:

------------------------Increasing $\Delta o$ ------------------------→

π-donor <weak π-donor<no π effects<π acceptor

Examples of representative ligands for these various classes are:
- **π-donor --- $I^-$**
- **weak π-donor ---$Cl^-$, $F^-$**
- **little or no π effects ---$H_2O$, $NH_3$**
- **π acceptor ---$PR_3$, CO**

# Inorganic Chemistry: Introduction to Coordination Chemistry

## Terms and Definitions for Chapter 6

*Sigma Bonding*

*Bond Order*

*Pi Bonding*

*Diatomic Molecules*

*Antibonding Orbital*

*Molecular Orbitals*

*Backbonding*

*Dynamic Synergism*

*Organophosphine Ligands*

*Zeigler-Natta Catalysts*

*Cone Angle*

*π-donor ligands*

*π-acceptor ligands*

*Alkene ligands*

*Wilkinson's Hydrogenation Catalyst*

*Linear Combination of Atomic Orbitals*

*Highest Occupied Molecular Orbitals (HOMO)*

*Lowest Unoccupied Molecular Orbitals (LUMO).*

# Inorganic Chemistry: Introduction to Coordination Chemistry

## Chapter 6 Problems and Exercises

1. Use MO theory to determine the bond order in each of the two following species: $[He_2]^+$ and $[He_2]^{2+}$. Does the MO scheme indicate that these species have viable bonds?

2. Construct the MO diagram for $O_2$, but only show the 2-p valence orbitals involved. Next, use the diagram to *rationalize* the trend in oxygen-oxygen bond distances for the following species:
$O_2$ = 121 pm
$[O_2]^+$ = 112 pm
$[O_2]^-$ = 134 pm
$[O_2]^{2-}$ = 149 pm
Which species is paramagnetic?

3. Draw the $a_{1g}$, the $t_{1u}$, and the $e_g$ and $t_{2g}$ sets of molecular orbitals that are expected for an octahedral metal complex. Make sure to show the overlaps between the metal and ligand based orbital portions.

4. Of the molecular orbitals drawn in problem #3 above, explain which are bonding and which are nonbonding.

5. Draw the scheme shown on page 163 that shows the MO diagram for a strong-field octahedral $[Co(NH_3)_6]^{3+}$ metal complex with a strong σ-donor ligand such as ammonia. Make sure that you label the $t_{2g}$ set as non-bonding. Make sure that you draw the box around the HOMO and LUMO orbitals that define the energy spread for $\Delta o$.

6. Draw the scheme shown on page 164 that shows the MO diagram for an octahedral metal complex with a common π-donor ligand such as chloride. (Use $[CrCl_6]^{3-}$) Make sure that you draw the box around the HOMO and LUMO orbitals that define the energy spread for $\Delta o$.

7. Draw the scheme shown on page 175 that shows the MO diagram for an octahedral metal complex with a common π-donor ligand such as carbon monoxide. (Use $[Cr(CO)_6]$) Make sure that you draw the box around the HOMO and LUMO orbitals that define the energy spread for $\Delta o$.

# Inorganic Chemistry: Introduction to Coordination Chemistry

8. Draw the synergistic bonding between a transition metal and a carbon monoxide ligand using the orbital overlap description.

9. Draw the synergistic bonding between a transition metal and a carbon monoxide ligand using the Lewis electron dot method showing the two resonance forms that are possible.

10. Why would iodide ion be lower on the spectrochemical series than the fluoride ion?

11. What kind of ligand would you expect the cyanide ion to be as far as sigma and pi interactions are concerned?

12. Why would water ion be lower on the spectrochemical series than the ammonia molecule? (hint—pi interactions)

13. Why would ammonia be lower on the spectrochemical series than the o-phenanthroline (phen) ligand.

14. Which would have the largest cone-angle… trimethyl phosphine or trimethyl phosphite?

15. Explain with drawings and discussion why Zeise's salt exhibits the structure with respect to the alkene ligand that it does.

# Chapter 7. Transition Metal Organometallics

The leading journal of the field, Organometallics, is an ACS publication and it uses this definition of "organometallics" in its Author Information section of its website.

"For the purposes of this journal, an 'organometallic' compound will be defined as one in which there is a bonding interaction (ionic or covalent, localized or delocalized) between one or more carbon atoms of an organic group or molecule and a main group, transition, lanthanide, or actinide metal atom (or atoms). Following longstanding tradition, organic derivatives of the metalloids (boron, silicon, germanium, arsenic, and tellurium) will be included in this definition. Furthermore, manuscripts dealing with metal-containing compounds that do not contain metal-carbon bonds will be considered as well if there is a close relationship between the subject matter and the principles and practice of organometallic chemistry. Such compounds may include, inter alia, representatives from the following classes: molecular metal hydrides; metal alkoxides, thiolates, amides, and phosphides; metal complexes containing organo-group 15 and 16 ligands; metal nitrosyls. Papers dealing with certain aspects of organophosphorus, organoselenium, and organosulfur chemistry also will be considered."

Organometallics is a huge area of chemistry that has come into its own in the last thirty years. Many organometallic compounds such as n-butyl lithium and the Grignard reagents (magnesium based) are essential building blocks in organic chemistry. Most organometallic compounds of the main-group metals are very reactive and many are extremely unstable. Grignard reagents are often air and water sensitive, and will easily decompose. Trivalent compounds such as $AlR_3$ and $GaR_3$ are electron-deficient (strong Lewis Acids) and are often ***pyrophoric***.

# Inorganic Chemistry: Introduction to Coordination Chemistry

***Transition metal organometallics*** is a large field that is growing rapidly. It is an advanced topic, but it may be interesting to show a range of compounds that are found in this field. The metal carbonyls are found throughout the transition series, and the first row transition metals form these ***binary metal carbonyls***:

[structures of binary metal carbonyls: $[V(CO)_6]^-$, $Cr(CO)_6$, $Mn_2(CO)_{10}$, $Fe(CO)_5$, $Co_2(CO)_8$, $Ni(CO)_4$]

Binary metal carbonyls are also called ***homoleptic compounds*** since they *only have one type of ligand* around the metal.

The series shown above has an interesting trend. All of these compounds show a total of eighteen valence electrons around the metal center.

When a metal compound has eighteen electrons around the metal, it is said to obey the ***eighteen-electron rule*** (similar to the octet rule), which allows the metal to achieve a noble gas configuration. The eighteen-electron rule is also called the ***noble gas formalism***, and the ***Effective Atomic Number (EAN) rule***.

Another interesting note about this series of metal carbonyls is the bimetallic structures that are seen with manganese and cobalt. Since the two metals have an odd number of electrons, they cannot form neutral compounds with carbon monoxide alone unless they form metal-metal bonds. In this fashion they are able to achieve eighteen electrons around each metal.

You should count electrons for each compound to confirm that they each have eighteen electrons. A neutral vanadium carbonyl [V(CO)$_6$] can't form an eighteen electron species because it would have only 17-electrons around the metal center, and it can't form a seven coordinate complex because of steric crowding. It is therefore very reactive, and it gains 18-electrons by becoming an anion [V(CO)$_6$]$^-$.

Electron counting will be discussed later in this chapter, and we'll work several examples. An example is Cr(CO)$_6$. Chromium is zero valent, and it therefore has a total of six valence electrons. Each carbonyl ligand donates two electrons for a total of twelve more electrons. Thus 6e$^-$'s + 12 e$^-$'s = 18 total.

Metal carbonyls are unusual in that the metal atom usually has a low oxidation state or has an oxidation state of zero. This is a consequence of the metal – carbonyl backbonding discussed in the last chapter, which stabilizes metal centers with larger numbers of d-electrons.

Carbon monoxide can also function as a bridging ligand, as can be seen for the dicobalt octacarbonyl compound (Co$_2$CO$_8$), or it can also bridge three metals.

**Modes of carbonyl bonding are:**
**Terminal        μ-CO              μ$_3$-CO**

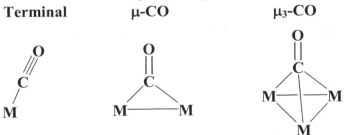

In the IR spectrum, free CO has a stretching mode at 2143 cm$^{-1}$. Absorption due to CO is strong and easily observed for metal carbonyls using infrared spectroscopy.

**The lower the value of the stretching frequency, the weaker the CO bond, and conversely, the greater the back-bonding and the greater the metal-carbon bond.**

The conclusions drawn from infrared data is supported by the structural data acquired from x-ray crystallography (and electron diffraction).

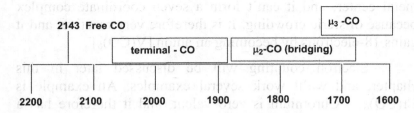

IR Stretching frequency in cm$^{-1}$

For free CO the carbon – oxygen bond length is 112.8 pm, whereas the terminal CO bond length in a metal carbonyl is typically 117 pm. A typical bridging CO bond length is longer, around 120 pm. <u>These longer bond lengths indicate stronger backbonding from the metal and a weaker C-O bond.</u> This experimental data strongly supports our molecular orbital approach to bonding for metal carbonyl compounds.

It is instructive to examine the data for an isoelectronic series of metal carbonyl species. For example, let's look at the following isoelectronic series of metal carbonyls and their CO stretching frequencies as well as their metal-carbon bond distances.

| **Complex** | **γ-CO/ cm-1** | **M-C (pm)** |
|---|---|---|
| $Ni(CO)_4$ | 2060 | 184 |
| $[Co(CO)_4]^-$ | 1890 | 175 |
| $[Fe(CO)_4]^{2-}$ | 1790 | 174 |

The data shows that as more negative charge is placed on the metal center going from $Ni^0$ to $Co^-$ to $Fe^{2-}$, then the backbonding increases. (This is because of the negative charge and the formal oxidation state of the metals, $Ni^0$, $Co^{-1}$, $Fe^{-2}$) This increased negative charge, and the subsequent greater backbonding, results in a lowering of the CO stretching frequency and a shortening of the metal-carbon bond.

Now, can you explain the data for the series above with respect to charge on the metal complex, backbonding, and the IR data?

| Complex | γ-CO/ cm-1 |
|---|---|
| $[Mn(CO)_6]^+$ | 2101 |
| $Cr(CO)_6$ | 1981 |
| $[V(CO)_6]^-$ | 1859 |

Metal carbonyls also have a tendency to form *metal clusters* that are nearly small chunks of metal surrounded by a carbon monoxide layer.

The metal clusters all exhibit metal-metal bonding. The metals share electron pairs, so in effect, each metal in a metal-metal bond is acting like a ligand to the other metal.

## Fluxional Molecules

The metal carbonyls are often very *fluxional molecules*, where the carbon monoxide ligands around the metal or metal centers are in motion and can vacillate between different geometries and can exchange between bridging and terminal sites.

An example of a geometric conversion has been discovered for iron pentacarbonyl, $Fe(CO)_5$, which undergoes what is called the ***Berry-Pseudorotation mechanism*** to convert between the trigonal bipyramid geometry and the square pyramidal geometry. This mechanism allows the CO ligands at the two different sites of the trigonal bipyramid (the axial and equatorial sites---see Chapter 2) to interconvert on the NMR time scale.

**Trigonal Bipyramid** ⇌ **Square Pyramid** ⇌ **Trigonal Bipyramid**

At room temperature the $^{13}C$ NMR shows only one peak, one kind of carbonyl, because the interconversion is so fast, but at colder temperatures, as seen with variable temperature NMR, the two different carbonyl ligands can be observed.

Metal carbonyl dimers and clusters often exhibit a fluxional process where terminal and bridging carbonyls interchange more rapidly with increasing temperature.

A commonly used example is that of dicobalt octacarbonyl, $Co_2(CO)_8$, which exhibits both a bridged form and an all-terminal form.

**Bridging** ⇌ **Terminal**

$Mn_2(CO)_{10}$ prefers the staggered form shown to the right, over the eclipsed form on the left which would exhibit far more steric hindrance.

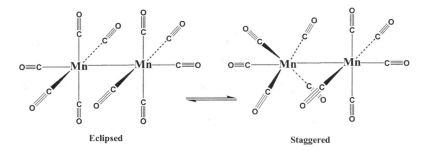

Eclipsed          Staggered

## Metallocenes

***Ferrocene*** was an important organometallic compound discovered in the 1950's due to the fact that it showed for the first time the synthesis of a metal-arene complex:

The cyclopentadienyl ring (cp⁻) is an anion and is aromatic (with six delocalized electrons). In electron counting, does ferrocene satisfy the eighteen-electron rule?

The cp⁻ rings are each six-electron donors giving a total of twelve electrons for the ligands. The iron atom has eight valence electrons, but since the molecule is neutral with no charge, the iron has to be formally in the $Fe^{+2}$ oxidation state. Thus, it has only six electrons. When added up, ferrocene has eighteen electrons around the iron center.

Many metal arene complexes have been synthesized since that time, including many *sandwich* type molecules similar to ferrocene.

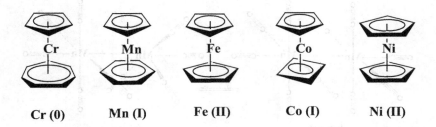

Cr (0)   Mn (I)   Fe (II)   Co (I)   Ni (II)

In a sandwich complex, the metal lies between two arene ligands. Complexes of the general type ($\eta^5$-Cp)$_2$M are called ***metallocenes***. (metallocenes do not *have* to be $\eta^5$-Cp) Can you determine if the metallocenes depicted above all obey the eighteen-electron rule? (see problem 17)

**Electron Counting Revisited**

Even though we covered two electron-counting methods in Chapter 5, the needs of transition metal organometallic chemistry are such that it is important to revisit the more important of the two methods, the **ICC** method. One of the simplest cases is for the binary metal carbonyl, chromium hexacarbonyl.

$$Cr(CO)_6$$

| | |
|---|---|
| Cr | 6e- |
| 6 terminal CO | 12e- |
| | 18e- |

This electron counting is done for the chromium center, and therefore the compound obeys the 18-electron rule. The formal oxidation state of chromium in this case is Cr°.

A more complicated case arises when a metal-metal bond is present, such as in the dimeric binary metal carbonyl compound, dimanganese decacarbonyl.

**$Mn_2(CO)_{10}$**

| | |
|---|---|
| Mn | 7e- |
| M-M bond | 1e- |
| 5 terminal CO | 10e- |
| | 18e- |

In this case the metal-metal bond plays an important role. The $Mn(CO)_5$ fragment acts as a ligand for the metal being counted, and becomes a one electron donor ligand through sharing. Each metal center has the same number of electrons: 7e⁻ from the metal, 1e⁻ from the other metal in the metal-metal bond, and 10e⁻ from the five terminal carbonyl ligands. The metal is formally in the $Mn^0$ oxidation state; the other Mn atom does not oxidize it.

Another example of a dimeric binary metal carbonyl is dicobalt octacarbonyl.

**$Co_2(CO)_8$**

| | |
|---|---|
| Co | 9e- |
| M-M bond | 1e- |
| 3 terminal CO | 6e- |
| 2 bridging CO | 2e- |
| | 18e- |

The difference between these two comes from the appearance of the bridging carbonyl binding mode. Since the carbon monoxide ligand is only using a pair of electrons to bond to the two metal centers, then, by convention, we assign only one electron to each metal. Again, in this case, the metal is formally in the $Co^0$ oxidation state; the other Co atom does not oxidize it, nor does the bridging carbonyls.

Continuing on, let's examine ferrocene. Ferrocene has no carbon monoxide ligands, but is synthesized by using an in-situ production of cyclopentadienyl anion (Cp⁻).

## Ferrocene

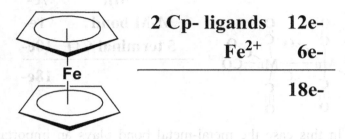

2 Cp- ligands    12e-
$Fe^{2+}$    6e-
_____
            18e-

For ferrocene, it is best to start with the ligands and remove them with their full valence shell electrons. Thus, we remove the ligand as Cp⁻, which is aromatic. The Cp⁻ anion is therefore a six-electron donor ligand, and this results in an oxidation of iron by plus one for each Cp⁻ ring. Even though iron is formally in the $Fe^{+2}$ oxidation state in ferrocene, it must be realized that this is a convention only. Ferrocene is non-polar, dissolves in organic solvents, and does not behave like an ionic compound. It is a molecular compound.

The following compound is similar to ferrocene, and can be synthesized from it, but it has a benzene ring as a ligand.

$$[(\eta^6-C_6H_6)(\eta^5-C_5H_5)Fe]^+$$

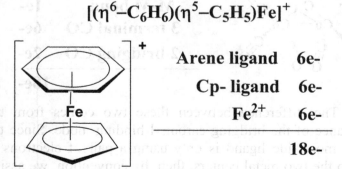

Arene ligand    6e-
Cp- ligand    6e-
$Fe^{2+}$    6e-
_____
            18e-

The complex has a Cp⁻ ring which is a six electron donor and a benzene (arene) ring which donates six-electrons. The overall charge on the complex is 1⁺, and since the Cp⁻ ring is negatively charged, the iron atom has to have an oxidation state of $Fe^{-2}$. Thus, the compound obeys the 18-electron rule.

Cycloheptatriene acts as an aromatic 6-electron donor by loss of a hydride (H⁻) to form the tropylium ion.

### Cycloheptatriene

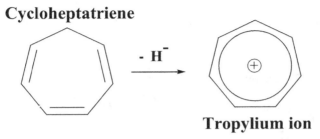

**Tropylium ion**

The *tropylium ion* reacts with transition metals in a similar fashion to benzene or cyclopentadienyl anion, except that it has a positive (1⁺) charge.

This metal complex has an overall charge of 1⁺, so the molybdenum has a formal oxidation state of $Mo^0$.

$$[(\eta^7-C_7H_7)Mo(CO)_3]^+$$

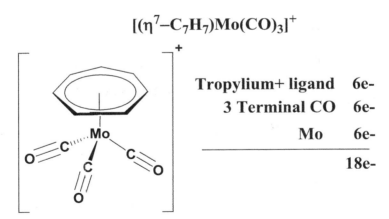

| | |
|---|---|
| Tropylium+ ligand | 6e- |
| 3 Terminal CO | 6e- |
| Mo | 6e- |
| | 18e- |

The cyclobutadiene molecule is not aromatic. However, in a beautiful twist of nature, cyclobutadiene can be made aromatic-where all the bond lengths are the same-by reacting it with a transition metal to form an organometallic compound.

In this case, the ligand is a four electron neutral donor ligand.

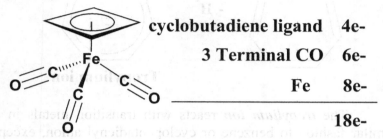

| cyclobutadiene ligand | 4e- |
| --- | --- |
| 3 Terminal CO | 6e- |
| Fe | 8e- |
| | 18e- |

The nitric oxide ligand is isoelectronic with carbon monoxide, but the major difference is that has a *positive charge*.

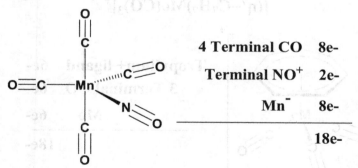

| 4 Terminal CO | 8e- |
| --- | --- |
| Terminal NO$^+$ | 2e- |
| Mn$^-$ | 8e- |
| | 18e- |

The interesting feature of the nitrosyl ligand is that it *decreases* the oxidation state of the metal.

Thus, the manganese metal center has an oxidation state of <u>minus one</u> (Mn$^-$)!

Vaska's complex undergoes oxidative addition to form six-coordinate octahedral complexes, some of which contain stable alkyl metal bonds.

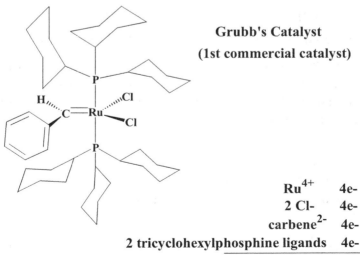

| | |
|---|---|
| triphenylphosphine ligands | 4e- |
| Terminal CO | 2e- |
| $CH_3^-$ | 2e- |
| $Cl^-$ | 2e- |
| $I^-$ | 2e- |
| $Ir^{3+}$ | 6e- |
| | 18e- |

In this case the triphenyl phosphine ligands are neutral, but the chloride, the iodide and the methyl groups are all anionic ligands. These three anionic ligands oxidize the iridium to $Ir^{3+}$, so that is the formal oxidation state of the metal.

Grubb's first commercial catalyst is very useful in organic chemistry due to its usefulness as a metathesis catalyst. To count the electrons one must pull off the carbene in its closed shell configuration. This would make the carbene an anion with a 2⁻ charge. Thus, formally, the Ru would have to be in the $Ru^{4+}$ oxidation state, and the trigonal bipyramidal complex would be a sixteen electron species.

**Grubb's Catalyst**
**(1st commercial catalyst)**

| | |
|---|---|
| $Ru^{4+}$ | 4e- |
| 2 Cl- | 4e- |
| carbene$^{2-}$ | 4e- |
| 2 tricyclohexylphosphine ligands | 4e- |
| | 16e- |

## Oxidative Addition and Reductive Elimination Reactions

A variety of industrially important catalytic cycles have as their basis a series of reactions involving an *oxidative addition reaction*, a rearrangement and then a *reductive elimination reaction*.

What is an oxidative addition reaction? An oxidative addition reaction is as the name suggests---a ligand (a general example is ligand A-B) reacts with a metal complex that is *coordinatively unsaturated* in such a manner that the metal is formally oxidized by +2 (ICC method of electron counting). An excellent example can be made by using one of the classical metal complexes, called *Vaska's complex*, in an oxidative addition reaction with HBr. Vaska's complex is a square-planar iridium(I) complex that is, of course, coordinatively unsaturated.

### Oxidative Addition Reaction

$(Ph)_3P$......Ir......$Cl$ / $C$ / $P(Ph)_3$ / $O$

**Vaska's Complex**
**Ir(I)**
**Square-Planar 16 e⁻ species**

→ HBr →

$(Ph)_3P$......Ir......$Br$ with H above and Cl below, $C$, $P(Ph)_3$, $O$

**Ir(III)**
**Octahedral 18 e⁻ species**

Oxidative addition reactions are more likely to occur in metals with low oxidation states compared to those with a higher oxidation state. Thus, a Rh(I) species is more likely to undergo an oxidative addition reaction compared to a Rh(III) species.

Oxidative addition reactions cannot occur with metals in their highest oxidation state. For example, a Mn(VII) species such as $MnO_4^-$ cannot undergo an oxidative addition reaction.

Ligand species that have a weaker A-B bond are more likely to undergo an oxidative addition reaction compared to those with a stronger A-B bond. Thus, a C-Cl ligand bond is more likely to undergo oxidative addition as compared to a C-C ligand bond.

A reductive elimination reaction is the microscopic reverse reaction of an oxidative addition reaction. In the reductive elimination reaction the ligands that are eliminated *must be cis* to each other. An example is given below, where the iridium(III) complex (coordinatively saturated) eliminates acetone as a product, and iridium is formally reduced to Ir(I), and becomes coordinatively unsaturated.

**Reductive Elimination Reaction**

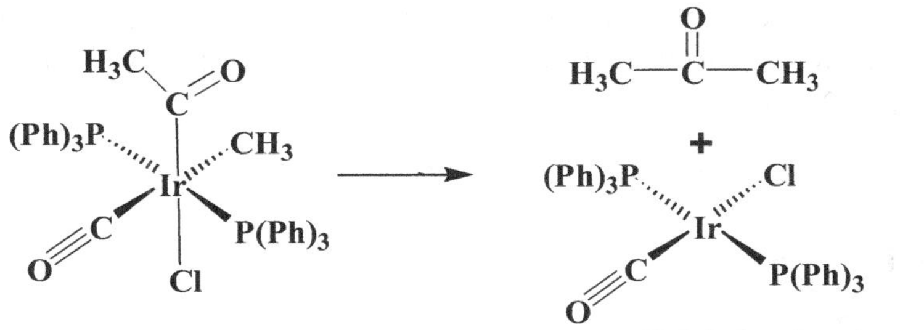

Ir(III)
Octahedral 18 e⁻ species

Vaska's Complex
Ir(I)
Square-Planar 16 e⁻ species

The reductive elimination step is a crucial step in many industrial catalytic processes (which explains the common use of rhodium) such as the Monsanto acetic acid process, the Tennessee Eastman acetic anhydride process, and in many hydrogenation reactions.

# Inorganic Chemistry: Introduction to Coordination Chemistry

## Metal-Metal Multiple Bonds

The existence of a ***quadruple metal-metal bond*** in inorganic compounds was first recognized in 1964 when the compound $[Re_2Cl_8]^{2-}$ was isolated.

Unfortunately, the complex was actually first synthesized ten years prior to this in the Soviet Union, but was characterized mistakenly as the Re(II) compound, $K_2ReCl_4$.

In a now classical paper, F. Albert Cotton assigned the correct formula and structure as being a one that contained a Re-Re quadruple bond due to the odd eclipsed structure uncovered by x-ray crystal analysis.

The compound consists of two $d^4$ $ReCl_4$ groups, which contain Re(III) metal centers, and a metal-metal bond distance of only 223 pm.

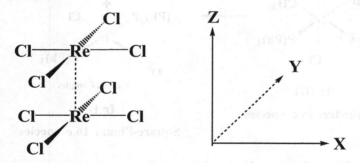

The scheme above shows that the exhibited geometry is that of two square-planar $ReCl_4$ units stacked on top of one another to form an eclipsed structure where the eight Cl ligands form a nearly perfect cube.

From an electrostatic and steric point of view, this compound should exhibit the staggered conformation.

Since the staggered conformation is not observed, then there must be unexplained bonding considerations.

The bonding can most easily be explained by considering the space orientation of the d orbitals. Each rhenium is slightly displaced above the center of a square planar array of four chloride ions. The metal $dx^2-y^2$ orbital has the appropriate symmetry to bond to the four chlorides.

The 1st bond between the metals uses the dz orbitals on each metal, to form a σ bonding molecular orbital.

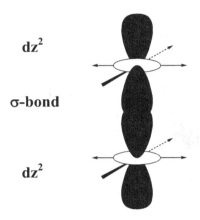

**$dz^2$**

**σ-bond**

**$dz^2$**

The 2nd and 3rd bonds between the metals use the dxz and the dyz orbitals on each metal, to form two π bonding molecular orbitals. Both π-bonding MO's are shown here.

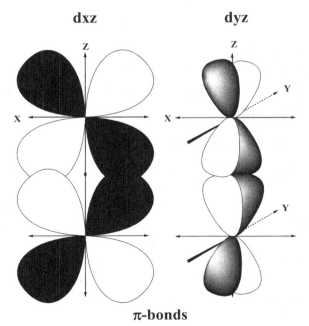

**π-bonds**

The 4$^{th}$ bond between the two metals is formed by the two dxy orbitals that are parallel to each other and form a bond not seen in organic chemistry called a *δ (delta) bond*. The "delta", with a "d", name comes from the fact that it looks like a d-orbital-shaped bond. The $dx^2$-$y^2$ orbital is used to form M-L bonds and is not involved in the M-M bond.

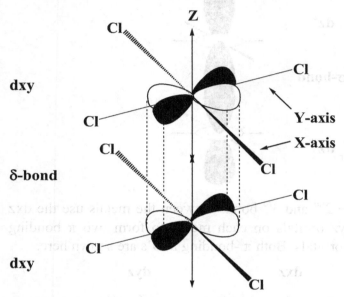

For overlap between the two dxy orbitals to be maximized in order to become a bonding interaction, the two ReCl$_4$ square planes <u>must be eclipsed</u> to each other. Despite the fact that this maximizes inter-atomic repulsions between the chloride ligands, the ability of the two metals to quadruple bond is the overriding stabilizing factor.

Perhaps the most interesting feature of this compound is the δ interaction in the quadruple bond. Because the δ orbital is only weakly bonding and the δ* orbital is only weakly antibonding a number of interesting chemical and spectroscopic consequences result. For example, the brilliant blue color of [Re$_2$Cl$_8$]$^{2-}$ is due to a δ ----> δ* electronic transition. Because of the weakness of the δ bond, the gain or loss of electrons has a relatively minor effect on the strength of the M-M bond.

The following scheme shows the summation of the bonds in the quadruple bond, with the delta bond being the highest energy molecular orbital.

**dxy**        δ-bond

**dxz, dyz**   π-bonds

**dz²**        σ-bond

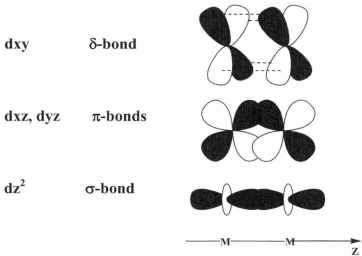

The MO energy diagram is shown below. The delta, pi, and sigma bonds have filled orbitals, but the antibonding orbitals are unfilled.

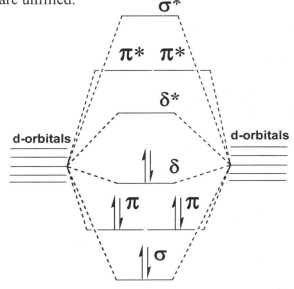

# Inorganic Chemistry: Introduction to Coordination Chemistry

**Terms and Definitions for Chapter 7:**
*Pyrophoric*

*Organometallics*

*Binary Metal Carbonyls*

*Homoleptic Compounds*

*Eighteen Electron Rule*
*Noble Gas Formalism*
*Effective Atomic Number (EAN)*

*Metal Clusters*

*Fluxional Molecules*

*Berry-Pseudorotation Mechanism*

*Ferrocene*

*Sandwich Type Molecules*

*Ziegler-Natta Polymerization*

*Oxidative Addition Reaction*
*Reductive Elimination Reaction*

*coordinatively unsaturated*

*Vaska's complex*

*Quadruple Metal-Metal Bond*

*δ (delta) bond*

*Tropylium Ion*

## Inorganic Chemistry: Introduction to Coordination Chemistry

**Chapter 7 Problems and Exercises**

1. Write the formulas for the neutral mononuclear metal carbonyl molecules formed by V, Cr, Fe, and Ni. Which ones satisfy the noble gas formalism?

2. Why are the simplest carbonyls of the metals Mn, Tc, Re, and Co, Rh, and Ir, polynuclear?

3. Explain, with necessary orbital diagrams, how CO, which has negligible donor properties toward simple acceptors such as $BF_3$ can form strong bonds to transition metal atoms.

4. In what ways can CO be bound to a metal atom?

5. Discuss and explain the trend in CO stretching frequencies in the series $V(CO)_6^-$, $Cr(CO)_6$, $Mn(CO)_6^+$.

6. Draw the structures of
   (a) $Fe_2(CO)_9$,
   (b) $Ru_3(CO)_{12}$
   (c) $Rh_4(CO)_{12}$

7. What kind of ligand would hydrogen be counted as in a transition metal complex? Count the electrons for hydridocobalt tetracarbonyl $[HCo(CO)_4]$ and verify that it obeys the 18-electron rule. What is the oxidation state of the cobalt?
What if you discover that the compound is acidic in water, would that change your electron counting formalism?

8. Confirm that the iron center in $H_2Fe(CO)_4$ obeys the 18-electron rule. What is the formal oxidation state of iron when using the ICC electron counting technique?

9. Confirm that the manganese center in $HMn(CO)_3(PPh_3)_2$ obeys the 18-electron rule. What is the formal oxidation state of manganese when using the ICC electron counting technique?

# Inorganic Chemistry: Introduction to Coordination Chemistry

10. By using the 18-electron rule show that the zero-valent metal carbonyls of the manganese Group (Mn, Tc, Re) should all exhibit a metal-metal bond in their $M_2(CO)_{10}$ compounds.

11. Confirm that the metal centers all obey the 18-electron rule in the following complexes. Some do not and have 16-electrons instead. Draw the structures.
    (a) cis-$Mo(CO)_4(pamp)_2$
    (b) $Fe(CO)_5$
    (c) di-$\mu$-carbonylbis(tricarbonylrhodium)(0)
    (d) $W(CO)_6$
    (e) $Mo(CO)_4$dipamp
    (f) $(\eta^5\text{-}Cp)_2Fe$
    (g) [RhH(triphenylphosphine)$_3$]
    (h) [Ir(CH$_3$)H(CO)Cl(triphenylphosphine)$_2$]
    (i) $^{99m}$[Tc(dmpe)$_3$]$^+$
    (j) $(\eta^5\text{-}Cp)Tc(CO)_3$
    (k) $(\eta^6\text{-}C_6H_6)_2Cr$
    (l) $(\eta^5\text{-}Cp)(CO)Fe(\mu\text{-}CO)_2Fe(CO)(\eta^5\text{-}Cp)$ this Homobimetallic compound contains a metal-metal single bond

12. Draw the Berry Pseudorotation mechanism for $Ru(CO)_5$ that makes all the carbonyl ligands interchange on the $^{13}C$ NMR timescale at room temperature and become equivalent.

13. The M-CO bonding is synergetic or involves synergism.
a. What is synergism?
b. Explain the following trend in the magnitude of $v_{C\text{-}O}$ IR frequencies:

free CO > terminal M-CO > $(\mu^2\text{-}CO)M_2$ > $(\mu^3\text{-}CO)M_3$

# Inorganic Chemistry: Introduction to Coordination Chemistry

14. For the following organometallic dimer complexes, determine the metal-metal bond order assuming that each complex obeys the 18-electron rule. Draw the structure of each complex and show how you counted the 18 electrons next to each structure.
    (a) di-μ-carbonylbis(tricarbonylcobalt)(0)
    (b) $[(\eta^5\text{-}C_5H_5)Mn(CO)_2]_2$
    (c) $[(\eta^5\text{-}C_5H_5)Mo(\mu\text{-}CO)_2]_2$

15. Draw the structure of the following compound and predict the bond order of any metal-metal bond in the $[Re_2Cl_6(\mu\text{-DIPHOS}]$ dimer.

16. Count the electrons for the metal sandwich compounds found on the cover, and verify that the oxidation states are correctly listed.

17. Draw the square planar $[RuCl(NO)(PPh_3)_2]$ complex. How many isomers are present? Verify that every isomer has sixteen or eighteen electrons around the metal center. What is the formal oxidation state of the Ru metal center?

18. Draw the tetrahedral $[Co(CO)_3(NO)]$ complex. Verify that it has sixteen or eighteen electrons around the metal center. What is the formal oxidation state of the Co metal center?

19. Draw the tetrahedral $[Fe(CO)_2(NO)_2]$ complex. Verify that it has sixteen or eighteen electrons around the metal center. What is the formal oxidation state of the Fe metal center?

20. Confirm that each different metal center in the following compound obeys the 18-electron rule. $[Fe_3(CO)_{12}]$

# Chapter 8. Bonding in Solids

We discussed earlier in chapter 1 that there were generally three types of bonding between atoms: (1) metallic, (2) covalent, and (3) ionic. It would be easy to also say that there are three types of inorganic homogeneous crystalline solids, but that would not be correct.

It is true that there are metallic solids and ionic solids, but covalent solids are found split into two categories. These categories are covalent molecular solids, and network covalent solids. What are the differences between the two?

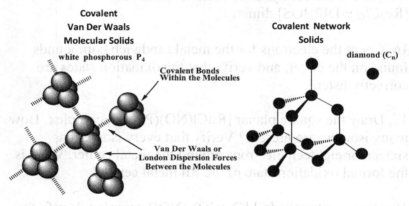

Molecular covalent solids consist of molecular compounds such as sulfur rings ($S_8$) or white phosphorous ($P_4$). The atoms of the molecules themselves are held together by covalent bonds, but the entire molecule is bonded to other molecules in the solid only by London Dispersion forces which are also know as Van der Waals forces. Thus, the bonding between individual molecules is very weak. On the other hand, network covalent solids are not molecular solids, and all the atoms are bound covalently. This is the case for solids such as diamond.

In non-metal covalent solids the electron sharing bonding between atoms is almost always between atoms within molecular units. Those bonding valence electrons are therefore almost always confined to the internuclear spaces between the bonding atoms. Electrons in covalent single bonds are not free to move throughout molecular solids.

In ionic solids the situation is even more severe, and valence electrons are confined to the individual ions themselves. On the other hand, metals definitely have electrons that are free to move throughout the solid.

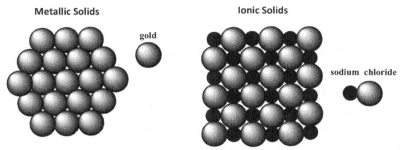

The three-dimensional crystal structures of ionic solids and metals can best be described by the way that they pack. The packing of the atoms in metallic solids can be described easily enough if it is done systematically.

There are four general types of packing arrangements, the crystal lattice, for the metallic elements. These are:

(1) Simple cubic (SC)  (2) Body-centered cubic (BCC)
(3) Hexagonal close-packed  (4) Cubic close-packed (FCC)

The first two examples are based on a crystal lattice that is not very efficient at packing, they are not very compact, and the *interstices*, the holes between atoms, are very large.

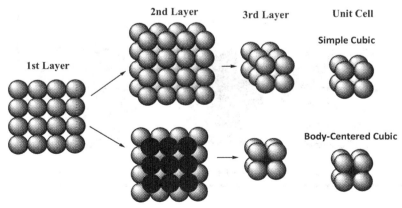

The simple cubic lattice does not pack very efficiently, and only one metal, polonium, packs in a simple cubic lattice. The body-centered cubic lattice is a little more efficient at packing, and the seconds layer of atoms are packed on top of the holes of the first. This has the effect of making the atoms pack more efficiently, and the metal becomes more dense. These two types of lattice are shown in the scheme above, along with the effect of layer addition of atoms, and the overall unit cell.

The last two types of lattice are based on an off-set hexagonal array of atoms as shown in the next scheme. Notice that this type of packing in the first layer is more efficient, and the interstitial hole is smaller than that seen in the simple cubic lattice above. The atoms are too close together for atoms of the next layer to fit above all the interstitial spaces, and only every other row of interstitial spaces can have an atom in the second layer fit over it.

For the hexagonal close-packed (HCP) lattice, the third layer of atoms fits exactly over the third layer. Thus for HCP, there are two distinct layers, and HCP is therefore also called an ABAB type lattice, where A is for the first layer, and B is for the second layer. Overall, the HCP lattice is far more efficient at packing than the simple cubic lattice, and the body-centered cubic lattice, and the interstitial holes are also smaller.

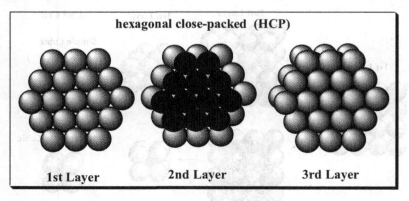

The last type of packing lattice is the cubic close-packed lattice, (CCP) and it has three distinct layers; it is an ABCABC type lattice.

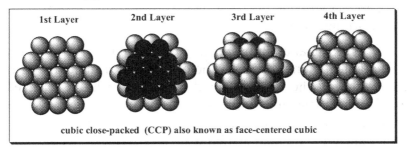

cubic close-packed (CCP) also known as face-centered cubic

The first and second layers are exactly the same as in the HCP case, but the third layer is over the holes that were not covered in the first layer by the second layer of atoms.. Then a fourth layer of atoms repeats the first layer. This is the most dense and efficient packing array of atoms, and can be seen in some of the most dense metallic elements.

The CCP lattice is also called the face-centered cubic lattice. The unit cells for the simple cubic, body-centered cubic and the face-centered cubic lattices are shown together in the next scheme.

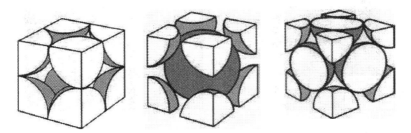

**Simple Cubic   Body-Centered Cubic   Face-Centered Cubic**

The unit cell is the smallest building block of a crystal, consisting of atoms, ions, or molecules, whose geometric arrangement defines a crystal's characteristic symmetry and whose repetition in space produces a crystal lattice.

It is interesting to note that the unit cells do not show all the atoms; the atoms on the edges and faces of the unit cell

# Inorganic Chemistry: Introduction to Coordination Chemistry

are only partial spheres. For the simple cubic lattice, the atoms shown at the edges are only 1/8 of an atom, so that the entire lattice, if it is a metal, would only consist of one atom, but yet the efficiency of the unit cell model is such that it shows the lattice arrangement of the whole crystal!

The metal atom occupancy for polonium in the simple cubic lattice is only at 52%. The other 48% of the unit cell is empty space. The body-centered cubic lattice has a unit cell that is more efficient, and the metal is at 68% occupancy. The hexagonal close-packed and the cubic close packed (or face centered cubic) are both at a high 74% metal atom occupancy.

If we do a survey of some of the metals in the periodic table as seen in the next picture, then we see an interesting situation at standard ambient temperature and pressure:

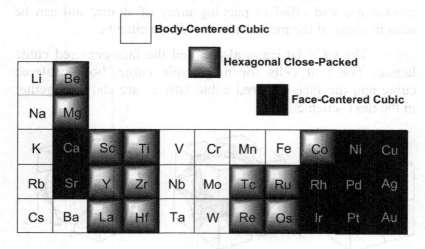

It seems that even though the trend is not perfect or well understood, in general, as the number of outer electrons increases across a period then the metals tend to prefer to pack in the order of BCC, then HCP, then FCC.

It is also interesting to note that the elements with the highest densities are osmium and iridium.

A theory of metallic bonding must account for and explain the key physical properties of metals. And, to be of any use, solid-state bonding models must explain the differences in conductors, semi-conductors, and insulators.

The most important property of metals is that they are electrical conductors in three dimensions at standard ambient temperature and pressure. Also, the theory should account for high thermal conductivity, high reflectivity (luster), and the physical working properties of malleability and ductility.

One of the simplest ideas that account for these properties is the "*electron sea*" model. In this simple model, valence electrons on metal atoms are free to move throughout the bulk solid, and even may leave the solid with electricity or photons (the photoelectron effect). The electron sea model says that metals have high thermal conductivity due to the flow of thermal energy carried by the movement of valence electrons through the solid.

The problem with the electron sea model is that it is a qualitative model, albeit one that is easy to understand, but it isn't quantitative. Measurements and observation can't be made or evaluated based on the electron sea model.

However, the molecular orbital theory that we have discussed earlier can be used and developed to explain the bonding in solid-state materials. This model is called "Band Theory".

## Band Theory

Molecular orbital theory is the most useful and comprehensive bonding theory that we have in chemistry, and it can be easily extended to explain the properties of metals and metallic bonding.

Metal crystals can be viewed as gigantic molecules whose size is determined only by the size of the metal crystal. Even a very tiny piece of a metal will contain tremendous numbers of atoms, and this fact has huge implications for the application of molecular orbital theory to metals.

In molecular orbital theory, as we discussed in Chapter 6, when two atoms such as hydrogen interact to form a bond, using a valence atomic orbital on each atom, then two molecular orbitals will be formed, one of lower energy that is bonding, and a higher energy antibonding orbital.

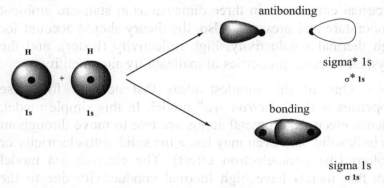

The third element in the periodic table, lithium, could also form a diatomic molecule $Li_2$, and a molecular orbital diagram can be drawn for the valence electrons in this simple diatomic molecule.

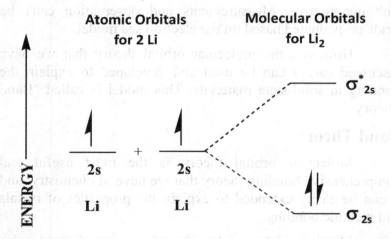

If there are more than two metal atoms that are bonded together, their atomic orbitals combine to form even more bonding and antibonding orbitals, and even non-bonding molecular orbitals.

The combination of all these molecular orbitals results in a continuous "Band" of molecular orbitals.

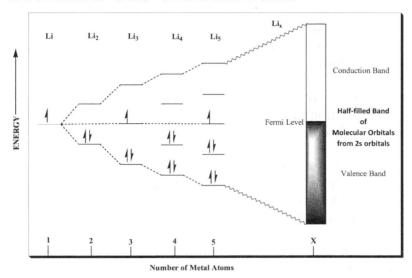

In a large crystal of a metal, like sodium, there are so many molecular orbitals formed that the spacing between the levels is so close that they essentially make up a continuum.

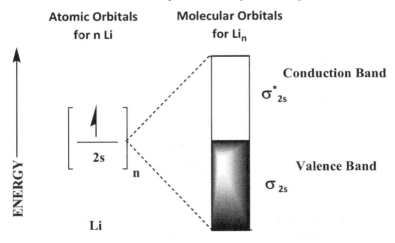

The lower energy MO's are the *Valence Band*, and the empty MO's are the *Conduction Band*. Electrons can flow easily from the valence band into the conduction band, and that is of course where electrical conductivity occurs, and electrons travel to transfer heat energy throughout the solid.

This easy flow of electrons into the conduction band is what gives metals the property of electrical conductivity in three dimensions, and the ability to also conduct heat.

The dividing line between the valence band level where electrons are found, and the overlapping conduction band is called the Fermi level.

Any time in a solid that there is efficient overlap of atomic orbitals in partially filled valence bands or there is overlap of low-lying empty bands with filled bands there is a chance for metallic-like conduction.

An example of this is sodium metal, which has the same number of valence electrons as lithium, but they are in the 3rd energy shell.

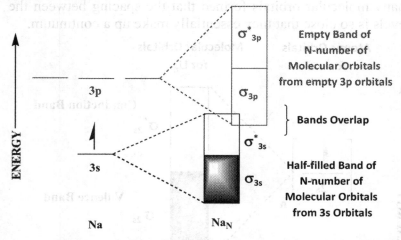

The example of sodium metal with a half-filled 3s orbital can be compared to magnesium which has a filled 3s orbital. In this case there is still an overlap with the low-lying 3p orbitals which are empty. These overlapping empty 3p orbitals serve as the conduction band.

# Inorganic Chemistry: Introduction to Coordination Chemistry

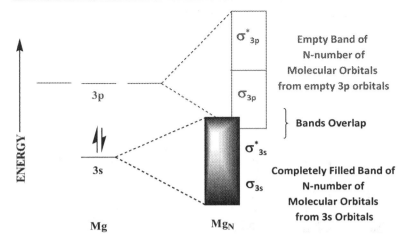

An interesting property of metals is that conductivity decreases with increasing temperature. This is because at higher temperatures there are more vibrations of metal atoms in the crystal lattice, and this causes higher resistance to electrical flow.

The reason that metal orbitals are able to function as conduction bands is that they are delocalized. The electron pairs are directed in typical covalent bonds and they occupy mainly the internuclear space between atoms.

In metals, the metal-based atomic orbitals form molecular orbitals that are delocalized throughout the surrounding atoms.

Remember, metals also are very large compared to the semi-metals and the non-metals; nonmetals are small and their valence electrons experience larger effective nuclear charges than do the valence electrons of the metals.

The Band Gap theory can also help explain the differences in the electrical conducting ability of metals, semi-metals and insulators.

In the elements that are semiconductors, the semi-metals, atoms are smaller than the metals and tend to form more covalent bonds, and this lowers the ability of the electrons to form delocalized molecular orbitals.

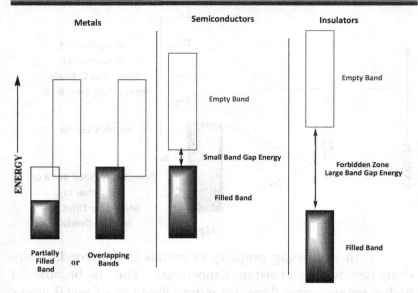

Instead of overlap between bands, the energy between the filled and empty molecular orbital bands is small, but apparent, and is called the *Band Gap Energy*. The band gap for an insulator, is high. Diamond is an insulator, and it has a band gap of 5.47 electron volts (eV). Another insulator is silicon carbide, and it has a band gap of 3.00 eV.

The most common semiconductor is silicon, and it has a band gap of 1.12eV. A smaller band gap is seen for a rarer and more expensive semiconducting element, the semi-metal germanium, which has a band gap of 0.66 eV.

The important characteristic physical property of semiconductors, like silicon or germanium, is that with an increase in temperature there is an increase their conductivity, because some electrons are able to overcome the band gap energy with increased temperature.

There are in general two different types of semiconductors; intrinsic semiconductors and extrinsic semiconductors. Intrinsic semiconductors are pure elements such as pure silicon, pure germanium, or pure gallium.

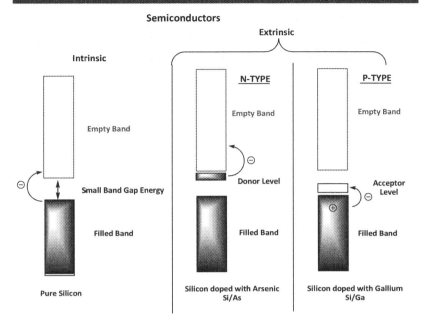

Extrinsic semiconductors are intrinsic semiconductor materials such as silicon that have been alloyed, or doped, with very low levels of an impurity element.

There are two types of extrinsic semiconductors, based on the dopants used.

N-type semiconductors are typically made from a pure Group IV element such as silicon that has been doped with a Group V element such as arsenic. The arsenic dopant atoms are richer in electrons, and these electrons make up a donor level that more easily contribute electrons to an empty conduction band. They are called N-type extrinsic semiconductors because they provide extra electrons (negative charge) to the substrate semiconductor.

The other type of extrinsic semiconductor is the P-type. The P-type semiconductors again have the basic Group IV semiconductor material but this time it has been doped with a Group III element such as gallium. A Group III element has fewer electrons than a Group IV element, so the dopant atoms act as an electrons "hole". The electron "hole" is like a positive charge in the semiconductor, and this positive charged hole can transmit electrical current. They are called P-type extrinsic semiconductors because they provide a positive hole (positive charge) to the substrate semiconductor.

Extrinsic semiconductors are intrinsic semiconductor materials such as silicon that have been alloyed, or doped, with very low levels of an impurity element.

There are two types of extrinsic semiconductors, based on the dopants used.

N-type semiconductors are typically made from a pure Group IV atom as such as silicon that has been doped with a Group V element such as arsenic. The arsenic dopant atoms are richer in electrons, and these electrons are up in donor level that more easily contribute electrons to an empty conduction band. They are called N-type extrinsic semiconductors because they provide extra electrons (negative charge) to the substrate semiconductor.

The other type of extrinsic semiconductor is the P-type. The P-type semiconductors again have the basic Group IV semiconductor material but this time it has been doped with a Group III element such as gallium. A Group III element has fewer electrons than a Group IV element, so the dopant atoms act as an electron "hole". The electron "hole" is like a positive charge in the semiconductor, and this positive

# Inorganic Chemistry: Introduction to Coordination Chemistry

**Terms and Definitions for Chapter 8:**

*Unit Cell*

*Simple Cubic (SC) Lattice*

*Body-Centered Cubic (BCC) Lattice*

*Hexagonal Close-Packed Lattice*

*Cubic Close-Packed (FCC) Lattice*

*Interstices*

*Interstitial space*

*Electron sea*

*Band Theory*

*Band Gap Energy*

*Filled Band*

*Valence Band*

*Conduction Band*

*Semiconductors*

*Metals*

*Ionic Compounds*

*Covalent Molecular Solids*

*Network Covalent Solids*

*Intrinsic Semiconductors*

*Extrinsic Semiconductors*

*Dopant*

*N-type Semiconductors*

*P-type Semiconductors*

Inorganic Chemistry: Introduction to Coordination Chemistry

## Chapter 8 Problems and Exercises

1. What is the type of bonding force holding all of the molecules of carbon dioxide together in solid dry ice?

2. Glass, which is a form of silicon dioxide, can best be described as a _____ type of solid.

3. Metallic solids are generally malleable and ductile whereas ionic solids are generally hard and brittle. Explain these observations.

4. Which type of metallic lattice is more efficient at packing?
(a) body-centered cubic or face-centered cubic
(b) face-centered cubic or simple cubic
(c) cubic close-packed or body-centered cubic

5. Draw a simple band scheme to distinguish a metallic conductor like a Group I metal such as sodium compared to a Group II metal such as calcium.

6. Draw a simple band scheme to distinguish a metallic conductor like a Group I metal such as sodium compared to an insulator such as diamond.

7. Draw a simple band scheme to distinguish between an N-type or P-type semiconductor.

8. Decide if the following systems are likely to be N-type or P-type semiconductors:
(a) arsenic doped germanium
(b) boron doped germanium

9. Which has a higher band gap energy?
(a) gallium or cesium  (b) sulfur or silicon
(c) phosphorous or aluminum  (d) zinc or germanium

10. The most important physical property of semiconductors is that their conductivity _____ with increasing temperature, whereas for metals their conductivity _____ with increasing temperature.

11. Both carbon in the diamond form and silicon exhibit the same tetrahedral like array of atoms. What is the difference in their bonding, and how can it be explained?

230

# Inorganic Chemistry: Introduction to Coordination Chemistry

12. Classify each of the following compounds as covalent molecular, covalent network, metallic or ionic, based only on their chemical formulas. (a) $NO_2$ (b) $CaF_2$ (c) $SF_4$ (d) Ag

13. Classify each of the following compounds based on their chemical formulas, and their physical properties.
(a) covalent molecular
(b) covalent network
(c) metallic
(d) ionic

| Compound type | Compound | Melting Point °C | Boiling Point °C | Electrical Conductor | |
|---|---|---|---|---|---|
| | | | | Solid | Liquid |
| | RbI | 642 | 1300 | no | yes |
| | $Se_8$ | 217 | 684 | poor | poor |
| | $MoF_6$ | 17 | 35 | no | no |
| | Pt | 1769 | 3827 | yes | yes |
| | BN | 3000 | -------- | no | no |
| | B | 2300 | 2550 | no | no |
| | $CeCl_3$ | 848 | 1727 | no | Yes |
| | $NO_3F$ | -175 | -45.9 | no | no |
| | Ti | 1675 | 3260 | yes | yes |
| | $TiCl_3$ | -25 | 136 | no | no |

14. Classify each of the following compounds as covalent molecular, covalent network, metallic or ionic, based only on their chemical formulas. (a) $PF_3$ (b) $PF_5$ (c) $SO_2F$ (d) Co (e) LiF (f) RbI (g) $LuCl_3$

15. How many metal atoms, regardless of type, would be in the simple cubic, the body-centered cubic, and the face-centered cubic unit cells?

16. Polonium crystallizes in a simple cubic lattice that has a unit cell with an edge length of 336 pm. (a) what is the mass of the unit cell? (b) What is the volume of the unit cell? (c) What would be the density of polonium?

17. Calculate the density of sodium metal if it is in a body-centered cubic unit cell with a length of 424 pm.

18. The atomic radius of iridium is 136 pm and it crystallizes in a face-centered cubic unit cell. Calculate the density of iridium.

19. Aluminum has a density of 2.69 g/cm$^3$, and the radius of an aluminum atom is 143 pm. Determine if it is true that aluminum crystallizes in a face-centered cubic unit cell.

Bonus questions

20. If chromium metal has an atomic radius of 124.9 pm, and a density of 7.20 g/cm$^3$, is the geometry of the chromium unit cell a body-centered cubic unit cell, or a face-centered cubic unit cell?

21. If copper metal has an atomic radius of 127.8 pm, and a density of 8.95 g/cm$^3$, is the geometry of the copper unit cell a body-centered cubic unit cell, or a face-centered cubic unit cell?

# History of transition metal/inorganic chemistry

**450BC Empedocles** says that all matter is formed from **four elements - earth, air, fire** and **water**

**400BC Democritus** proposes that matter is actually composed of tiny indivisible particles, *atomos*

**1661 Boyle's "The Sceptical Chymist"** is published, introducing concepts of **elements**, **acids** and **alkalis**, and refuting many earlier claims by Paracelsus and ancient philosophers

**1669 Hennig Brandt** discovers **phosphorus** by distilling urine

**1704 H. Diesbach** creates **"Prussian Blue"** by accident, now known as $Fe_4[Fe(CN_6)]_3 \cdot H_2O$

**1766 Hydrogen** discovered by **Henry Cavendish**

**1774** Joseph **Priestley** discovers "dephlogisticated air", which **Lavoisier** renames **"oxygene"**

**1774-89 Lavoisier** proposes **Law of Conservation of Mass**

**1798 Tassaert** discovers $CoCl_3 \times 6NH_3$, one of the first known **coordination compounds**

**1803 Dalton** publishes table of **comparative atomic weights**

**1805 Gay-Lussac** proves that **water** is composed of two parts hydrogen to one part oxygen

**1808 Dalton** publishes his **atomic theory**

**1813/14 Berzelius** develops the **chemical symbols** and **formulas** used today

# Inorganic Chemistry: Introduction to Coordination Chemistry

**1817 Gmelin's Handbook** of inorganic chemistry

**1822 Gmelin** prepares **cobalt ammonate oxalates**

**1827 Zeise's salt** discovered, **$K^+[(C_2H_4)PtCl_4]$**, the first olefin complex

**1851 Genth, Claudet, Fremy** prepare $CoCl_3 \cdot 6NH_3$, $CoCl_3 \cdot 5NH_3$, etc.

**1852 Frankland** proposes the concept of **Valence** (all atoms have a fixed valence)

**1869 Blomstrand** develops **"Chain theory"** of cobalt ammonates

**1869 Dmitri Mendeleev** publishes his first periodic table with 63 known elements

**1884 Arrhenius** proposes his **electrolytic** theory of **ions**

**1890 Mond** and **Bertholet** prepare **$Ni(CO)_4$** and **$Fe(CO)_5$**

**1892 Alfred Werner's dream** about **coordination compounds**

**1897 J. J. Thomson** discovers that **electrons** (cathode rays) are negatively charged particles with very tiny mass

**1898 Marie and Pierre Curie** discover radium and polonium

**1900 Max Planck** proposes **Quantum Theory of electromagnetic radiation**

**1900** Francois **Grignard** discovers magnesium **Grignard reagents**

**1902 G.N. Lewis** utilizes the first primitive **Lewis electron dot structures** in class (<u>cubic structures</u>)

**1905 Alfred Werner's** paper on **coordination chemistry** that leads to **Nobel Prize**

**1905 Einstein** explains the **photoelectric effect of metals**, and the **wave /particle duality of light**

**1907 Werner** synthesizes both **isomers of $CoCl_3 \cdot 4NH_3$**

**1908 Fritz Haber** reveals the **Haber Process** for producing ammonia from nitrogen and hydrogen

**1909** The **pH scale of acidity** is devised by **Soren Sorensen**

**1911 Optical isomers of *cis*-$[CoCl(NH_3)(en)_2]X_2$ resolved by Werner**

**1912 Max von Laue** studies **X-ray diffraction** patterns and shows that crystals are repeating units of atoms

**1913 Bohr model of the atom** published (proposes the quantization of the energies of electrons)

**1913 Werner wins first Nobel Prize** given to an **inorganic chemist**

**1913/14** The **Periodic Table** in its **present form** was drawn up by **Henry Moseley**

**1919 Rutherford** discovers that elements can be **transmutated**

**1923 G.N. Lewis** proposes **Lewis Acid/Base Theory and uses Lewis Electron Dot Structures**

**1925 Fischer-Tropsch process** is developed

**1926 Shrödinger proposes the quantum-mechanical atom** (electrons in orbitals about nucleus; electron spectroscopy explained as transitions among orbitals)

**1927 Lewis ideas** applied to electronic interpretation of coordination compounds by **Sidgwick**

**1929 Crystal Field Theory** proposed by **Bethe**

**1930 Ziegler** and Gilman simplify organolithium preparation, using ether cleavage and alkyl halide metallation respectively

**1932 First application of Crystal Field Theory by Van Vleck**

**1935 Van Vleck combines Crystal Field Theory with Molecular Orbital Theory** to produce what is now called the **Ligand Field Theory**

**1937** Discovery of **Technetium** by **Emilio Segre**

**1939 Linus Pauling** publishes "The Nature of the Chemical Bond" which is in essence a summing up of seven papers of his published between 1931 and 1933

**1939-40 Sidgwick proposes VSEPR Theory**

**1940** McMillan and Abelson produce the first transuranic element, neptunium

**1941 Plutonium** synthesized by **McMillan** and **Glenn Seaborg**

**1951 Kealy, Pauson, Miller** obtain **ferrocene $(C_5H_5)_2Fe$** first sandwich complex is discovered

**1951 Orgel, Pauling, and Zeiss** describe the **bonding in metal carbonyls**
**1952 Henry Taube** describes classical **Inner-Sphere and Outer-Sphere electron transfer processes**

**1953** Karl **Ziegler** develops the first catalyst to convert monomers into polymers

**1954 Linus Pauling** wins the **Nobel Prize** for Chemistry for his work on chemical bonding

**1955 Ziegler and Natta** develop **olefin polymerization** at low pressure using mixed metal catalysts (transition metal halide / $AlR_3$)

**1956** Review by **Moffit** and **Ballhausen** leads to acceptance of Crystal Field Theory by inorganic chemists

**1958** The x-ray structure of **[CpMo(CO)₃]₂** reveals a **metal-metal bond**

**1959 Shaw and Chatt** describe an **Oxidative-Addition reaction**

**1961 Dorothy Hodgkin** works out the structure of **vitamin B12** using a computer, which shows a cobalt-carbon bond

**1961 Heck and Breslow** determine the cobalt-carbonyl catalyzed **hydroformylation** pathway

**1961** L. **Vaska** discovers that **(PPh₃)₂Ir(CO)Cl** reversibly binds $O_2$ (**Vaska's Complex**)

**1962 Xenon hexafluoroplatinate**, the first compound of an inert gas, synthesized by Neil **Bartlett**

**1963** Karl **Ziegler** and Giulio **Natta** awarded Nobel Prize for chemistry of high polymers

**1964 Fischer** isolates the first **metal-carbene** complex

**1965 Wilkinson** and Coffey, discover **(PPh₃)₃RhCl** (**Wilkinson's catalyst**) as a **homogeneous hydrogenation catalyst** for the hydrogenation of alkenes

**1965 Cotton** describes the **first quadruple metal-metal bond (delta bond)** in $[Re_2Cl_8]^{2-}$

**1965 Rosenberg** discovers that platinum compounds inhibit cell division, which leads to discovery of **cis-platin** as an anti-cancer drug

**1968** First **organouranium** compound, $(C_8H_8)_2U$

**1971 Yves Chauvin** was able to explain in detail how metatheses reactions function and what types of metal compound act as catalysts in the reactions

**1973 Fischer and Wilkinson** win the **Nobel Prize** in Chemistry for synthesis and discovery of organometallic compounds (sandwich type)

**1974 Cotton** and co-workers discover the first **agostic complex** with a metal---H-C interaction

**1979** The **Monsanto acetic acid process** from CO and methanol developed in the 1970's is described.

**1980 E.A. Deutsch** discovers that cationic complexes of $[^{99m}Tc(diars)_2X_2]+$ and $[^{99m}Tc(dmpe)_2Cl_2]+$ exhibit **myocardial uptake** ---thrusts **Tc** into **nuclear medicine** prominence

**1982 Monsanto asymmetric hydrogenation process (Knowles** etc.)

**1983 Nobel Prize** was awarded to **Henry Taube** for his work on the mechanisms of **electron transfer reactions** in metal complexes (Inner sphere and outer sphere)

**1990 Richard Schrock** produces the first efficient metal-compound catalyst for metathesis with metal carbenes

**1992 Robert Grubbs** develops a better metathesis catalyst which is stable in air. Commercial Ru catalysts have found many applications in organic chemistry (**Grubbs catalyst**)

**2001 Nobel Prize** was awarded to **Knowles, Noyori, Sharpless**, for work on **chirally catalyzed hydrogenation reactions** (Knowles & Noyori) and on chirally catalyzed oxidation reactions (Sharpless)

**2005 Nobel Prize** was awarded to **Yves Chauvin, Richard Schrock,** and **Robert Grubbs** for **metathesis catalysts**

## 1990-2000's
New generations of catalysts, supramolecular assemblies, molecular wires, and bioinorganic systems

# Chapter 1 Answers to Problems and Exercises

1. (a) magnesium bromide (b) calcium oxide (c) sodium sulfide (d) potassium nitrate (e) rubidium bromide (f) lithium fluoride (g) scandium chloride (h) zinc sulfide (i) aluminum bromide (j) cadmium iodide (k) strontium sulfate (l) magnesium phosphide

2. (a) $MgCl_2$ (b) $Na_2O$ (c) $K_2S$ (d) $Rb_3(AsO_4)$ (e) $CaCO_4$ (f) $Cs_2SO_4$ (g) $Al_2O_3$ (h) $BaSO_4$ (i) $Sr(NO_3)_2$ (j) $Li_3PO_4$ (k) $Al(OH)_3$ (l) $KMnO_4$ (m) $ZnO$ (n) $Sc_2(SO_3)_3$ (o) $CsCN$ (p) $Na_2CO_3$ (q) $K_2CrO_4$ (r) $NaNO_2$ (s) $NH_4Cl$ (t) $NaNO_3$ (u) $NH_4(C_2H_3O_2)$ (v) $(NH_4)_2SO_4$ (w) $(NH_4)_3PO_4$ (x) $Mg_3N_2$ (y) $NaClO_4$ (z) $Ca_2ClO_3$

3. (a) tin(II) fluoride (b) cobalt(III) oxide (c) tin(IV) fluoride (d) iron(II) oxide (e) iron(III) perchlorate (f) chromium(III) sulfate (g) iridium(III) nitrate (h) molybdenum(II) oxide (i) mercury(II) oxide (j) manganese(IV) oxide (k) lead(IV) nitrate (l) cobalt(II) bromide (m) titanium(IV) chloride (n) copper(II) nitrite (o) copper(I) cyanide (p) chromium(III) carbonate (q) palladium(II) chloride (r) lead(II) bicarbonate

4. (a) $FeCl_2$ (b) $Co_2O_3$ (c) $Cr_2O_3$ (d) $Fe(OH)_3$ (e) $TiBr_4$ (f) $CuSO_4$ (g) $Cr(CN)_3$ (h) $RhCl_3$ (i) $TcS_2$ (j) $RuI_3$

5. (a) $PbO_2$ (b) $CrCl_3$ (c) $Cu_2SO_4$ (d) $SnCl_2$ (e) $HgNO_3$ (f) $Sn(OAC)_4$ (g) $Pb(Cr_2O_7)$ (h) $Au(CN)_3$ (i) $Co_2(SO_3)_3$ (j) $Au_2SO_4$ (k) $HgBr_2$

6. (a) carbon monoxide (b) carbon dioxide (c) sulfur trioxide (d) diphosphorous pentoxide (e) tetraphosphorous decaoxide (f) nitric oxide (g) ammonia (h) phosphorous pentachloride (i) phosphine (j) sulfur tetrafluoride (k) xenon tetrafluoride (l) oxygen difluoride

7. (a) $XeF_2$ (b) $SO_2$ (c) $PCl_5$ (d) $B_2O_3$ (e) $N_2O_4$ (f) $SiBr_4$ (g) $BF_3$ (h) $NO$ (i) $N_2O$ (j) $NH_3$ (k) $PH_3$ (l) $SiO_2$ (m) $Cl_2O$ (n) $S_2Cl_2$ (o) $PCl_3$

# Inorganic Chemistry: Introduction to Coordination Chemistry

8. 2=linear or bent
   4= square planar or tetrahedral
   6= octahedral or hexagonal planar or trigonal prismatic

9. (d) 3

10. (a) $[Cr(NH_3)_4Br_2]Br$
    (b) $[Cr(NH_3)_4Br_2]Br \rightarrow [Cr(NH_3)_4Br_2]^+ + Br^-$

11. (a) six (b) octahedral (c) $Fe^{3+}$

12. (a) six (b) octahedral (c) $Mn^{3+}$

13. (a) six (b) octahedral (c) $Co^{3+}$

14. (a) six (b) octahedral (c) $Co^{3+}$

15.

| Compound | Ionic nitrites | Isomers |
|---|---|---|
| (1) $[Co(NH_3)_6](NO_2)_3$ | three | 1 |
| (2) $[Co(NH_3)_5NO_2](NO_2)_2$ | two | 1 |
| (3) $[Co(NH_3)_4(NO_2)_2]NO_2$ | one | 2 (cis and trans) |
| (4) $[Co(NH_3)_3(NO_2)_3]$ | 0 | 2 (fac and mer) |
| (5) $K[Co(NH_3)_2(NO_2)_4]$ | 0 | 2 (cis and trans) |
| (6) $K_2[Co(NH_3)(NO_2)_5]$ | 0 | 1 |
| (7) $K_3[Co(NO_2)_6]$ | 0 | 1 |

16.

| Compound | Ionic chlorides | Isomers |
|---|---|---|
| (1) $[Fe(H_2O)_6]Cl_2$ | two | 1 |
| (2) $[Fe(H_2O)_5Cl]Cl$ | one | 1 |
| (3) $[Fe(H_2O)_4Cl_2]$ | 0 | 2 (cis and trans) |
| (4) $NH_4[Fe(H_2O)_3Cl_3]$ | 0 | 2 (fac and mer) |
| (5) $(NH_4)_2[Fe(H_2O)_2Cl_4]$ | 0 | 2 (cis and trans) |
| (6) $(NH_4)_3[Fe(H_2O)Cl_5]$ | 0 | 1 |
| (7) $(NH_4)_4[FeCl_6]$ | 0 | 1 |

**17.**
**Compound**
(1) $[Pt(NH_3)_4](SCN)_2$
(2) $[Pt(NH_3)_3(SCN)]SCN$
(3) $[Pt(NH_3)_2(SCN)_2]$
(4) $NH_4[Pt(NH_3)(SCN)_3]$
(5) $(NH_4)_2[Pt(SCN)_4]$

**18. (a)**
(1) $[Pd(P\emptyset_3)_4]Cl_2$
(2) $[Pd(P\emptyset_3)_3Cl]Cl$
(3) $[Pd(P\emptyset_3)_2Cl_2]$
(4) $NH_4[Pd(P\emptyset_3)Cl_3]$
(5) $(NH_4)_2[PdCl_4]$

**(b) Compounds (1) and (5) would have the greatest conductivity in solution because (1) is a 1:2 electrolyte, and (5) is a 2:1 electrolyte. Each produce three equivalents of ions, the highest number for this series of compounds.**
**(c)** $[Pd(P\emptyset_3)_2Cl_2]$, two isomers, cis and trans

**19.** Three ions; 2 $Cl^-$ ions and $[Co(NH_3)_5Cl]^{2+}$

**20.** Five ions; 4 $K^+$ ions and $[FeCl_6]^{4-}$

**21.** Three moles of bromide ion

**22.** 1.96g x 1mol/267.28g = $7.33 \times 10^{-3}$ mol $[Co(NH_3)_6]Cl_3$
$7.33 \times 10^{-3}$ mol $[Co(NH_3)_6]Cl_3$ x 3mol $Cl^-$/mol $[Co(NH_3)_6]Cl_3$
= 0.0220 mol $Cl^-$
0.0220 mol $Cl^-$ x 143.32 g AgCl/1mol AgCl = **3.15 g AgCl**

**23.** 2.54g x 1mol/949.07g = $2.676 \times 10^{-3}$ mol $Ba_3[FeCl_6]_2$
$2.676 \times 10^{-3}$ mol $Ba_3[FeCl_6]_2$ x 3mol $Ba^{2+}$/mol $Ba_3[FeCl_6]_2$ = $8.03 \times 10^{-3}$ mol $Ba^{2+}$
$8.03 \times 10^{-3}$ mol $Ba^{2+}$ x 233.39 g $BaSO_4$/mol = **1.87 g $BaSO_4$**

**24.** 5g x 1mol/366.63g = 0.0136 mol $[Co(NH_3)_4Br_2]Br$
0.0136 mol $Br^-$/0.100 L = **0.136 M $Br^-$**

25. One $[Rh(NH_3)_5Cl]^{2+}$ ion and two $Cl^-$ ions

26. Three $NH_4^+$ ions and a $[FeCl_6]^{3-}$ ion

27. Two isomers for square planar, and one isomer for tetrahedral:

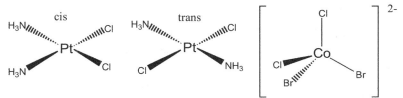

28.
(a) $Os = [Xe]6s^2 5d^6 4f^{14}$
(b) $Co = [Ar]4s^2 3d^7$
(c) $Ni = [Ar]4s^2 3d^8$
(d) $Ru = [Kr]5s^1 4d^7$
(e) $Cu = [Ar]4s^1 3d^{10}$
(f) $Re = [Kr]5s^2 4d^5$

29.
(a) $Zr = +4$
(b) $Ta = +5$
(c) $Mn = +7$
(d) $Nb = +5$
(e) $Tc = +7$
(f) $Y = +3$

30.
    (a) $[Ar] 3d^6 = Fe^{2+}, Co^{3+}$
    (b) $[Ar] 3d^5 = Mn^{2+}, Fe^{3+}$
    (c) $[Ar] 3d^{10} = Cu^+, Zn^{2+}$
    (d) $[Ar] 3d^8 = Ni^{2+}, Co^+$

31. (a) $Pt^{2+} = 5d^8$ (b) $Ni^{2+} = 3d^8$ (c) $Co^{3+} = 3d^6$ (d) $Ir^{3+} = 5d^6$ (e) $Ti^{4+} = [Ar]$ (f) $Zr^{4+} = [Kr]$ (g) $Cu^{2+} = 3d^9$ (h) $Ag^+ = 4d^9$ (i) $Cr^{3+} = 3d^3$ (j) $Mo^{3+} = 4d^3$ (k) $Mn^{3+} = 3d^4$ (l) $Tc^{3+} = 4d^4$

# Inorganic Chemistry: Introduction to Coordination Chemistry

**32.** six

**33.** $Ir^{3+}$

**34.** (c) the lanthanide contraction.

**35.**

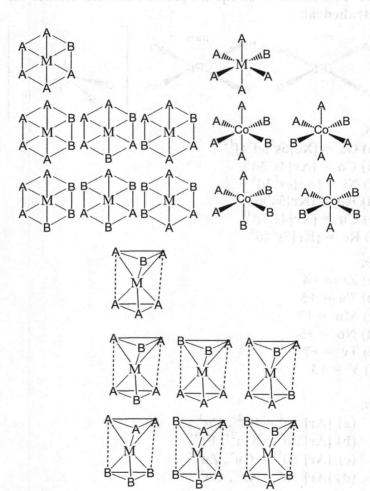

**Advanced**
**36. (a)**

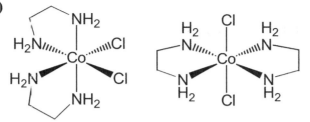

**(b)**

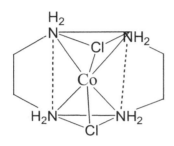 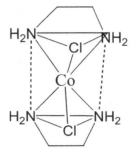

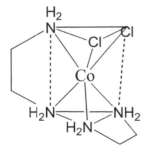 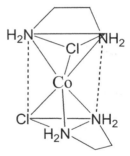

**(c)**

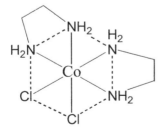

# Chapter 2 Answers to Problems and Exercises

**1.**
(A) $O_3$  (C) $CO_3^{2-}$  (D) $NO_2^-$

**2.** three

**3.** One

**4.**  $\quad$ C= -2  S= +2  N= -1
$\quad\quad$ [::C=S=N::]$^-$

**5.** $\quad$ N= -1  C= 0  S= 0
$\quad\quad$ [::N=C=S::]$^-$

**6.** (A) NO

**7.** Three

**8.** $Al^{3+}$, $BF_3$

**9.** Tetrahedral

**10.** Linear

**11.** Trigonal Planar

**12.** Slightly less than 109.5°

**13.** Trigonal Planar

**14.** 90° and 120°

**15. (A) COS**

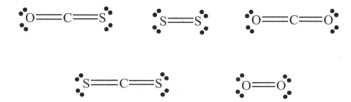

**16. (A) CH₂F₂**

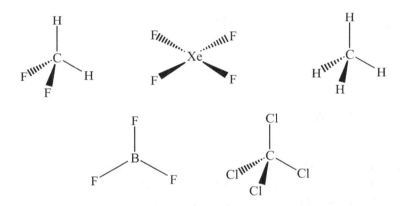

**17. (B) XeF₂**

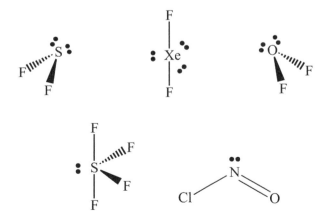

## 18.

(a) $Fe^{2+} = 3d^6$, $Fe^{3+} = 3d^5$, $\underline{Sc^{3+} = [Ar]}$, $Co^{3+} = 3d^6$

(b) $Tl^+ = 6s^2 5d^{10}$, $\underline{Te^{2-} = [Xe]}$, $Cr^{3+} = 3d^3$

(c) $Pu^{4+} = 5f^4$, $\underline{Ce^{4+} = [Xe]}$, $Tl^{3+} = 5d^{10}$

(d) $\underline{Ba^{2+} = [Xe]}$, $Pt^{2+} = 5d^8$, $Mn^{2+} = 3d^5$

## 19.

$[:C\equiv N:]^-$    $:C\equiv O:$    $PH_3$ (with lone pair on P, three H bonds)

$PF_3$ (with lone pair on P, three F bonds)    $[:S=C=N:]^-$

## 20.
Fluorine (a second row element with no possible d-orbital involvement) cannot have an expanded valence shell that holds more than eight electrons.

# Inorganic Chemistry: Introduction to Coordination Chemistry

## Key for Chapter 2 Practice Quiz

1. (e) all contain polar bonds
2. (e) none contain polar bonds
3. (e) none have a dipole moment
4. (e) $NH_3$
5. Tetrahedral = (b) $H_2O$
6. Trigonal Bipyramidal = (c) $PCl_5$
7. Octahedral = (d) $XeF_4$
8. Trigonal Planar = (e) $NO_2^-$
9. Linear = (a) $CO_2$
10. (a) true
11. Tetrahedral = (a) $CH_4$
12. Trigonal Bipyramidal = (c) $PCl_5$
13. Square Planar = (d) $XeF_4$
14. Bent = (b) $H_2O$
15. See-Saw = (e) $SF_4^-$
16. (c) tetrahedral
17. (a) pyramidal
18. (a) The ammonia molecule has tetrahedral molecular geometry.
19. (b) trigonal planar
20. (c) tetrahedral
21. (a) a single covalent bond.
22. (b) two
23. (a) one
24. (b) $O_3$
25. (b) two
26. (d) F-I
27. (e) $SbCl_3$
28. (d) $O_2$

# Chapter 3 Answers to Problems and Exercises

**1.**

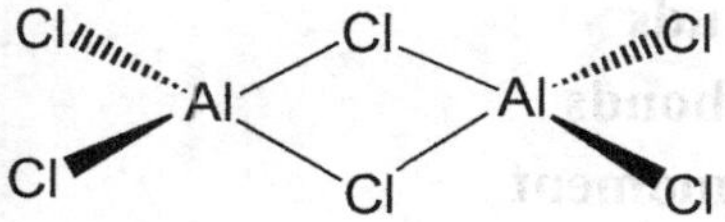

$$\text{Cl}_{\text{'''''}}\text{Al}\begin{matrix}\text{Cl}\\ \\ \text{Cl}\end{matrix}\text{Al}\begin{matrix}\text{''''''Cl}\\ \\ \text{Cl}\end{matrix}$$

$\text{Cl}$ $\text{Cl}$

**2.**
(A) ethylenediamine = bidentate
(B) aqua = monodentate
(C) cyano = monodentate
(D) carbonyl = monodentate
(E) ammine = monodentate

**3.** The metal is a <u>Lewis Acid</u> and the ligands are Lewis Bases.

**4.** (a) chloride (d) $CH_3^-$ (e) iodide

**5.** a Lewis acid

**6.**
   (a) I-
   (b) $S^{2-}$
   (c) $AsH_3$
   (d) dppe
   (e) thiocyano
   (f) dtc (dithiocarbamate)
   (g) pamp

**7.** (a) $Cu^{2+}$ (b) $Ni^{2+}$ (c) $Cr^{3+}$

**8.** (a) diphos (b) EDTA
   (c) acac (d) diop
   (e) porphyrin (f) dmpe
   (g) cyclam

**9.**
  (a) ammonia = classical
  (b) water = classical
  (c) chloride = classical
  (d) EDTA = classical
  (e) en = classical
  (f) diphos = nonclassical
  (g) acac = classical
  (h) DuPhos = nonclassical
  (i) porphyrin = classical
  (j) dipamp = nonclassical
  (k) carbon monoxide = nonclassical
  (l) methylamine = classical

**10.**

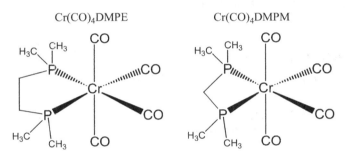

The DMPE ligand forms a five-membered chelate ring, but the DMPM ligand forms a more strained four-membered ring.

**11.**
(a) $[Pt(NH_3)_4Cl_2]SO_4$
dichlorotetraammine platinum(IV) sulfate

(b) $K_3[Mo(CN)_6F_2]$
potassium difluorohexacyanomolybdenate(V)

(c) $K[Co(EDTA)]$
potassium ethylenediamminetetracetatocobaltate(III)

(d) $[Co(NH_3)_3(NO_2)_3]$
trinitrotriammine cobalt(III)

## 12.
(a) $[Pt(NH_3)_6]Cl_4$
hexaammineplatinum(IV) chloride

(b) $[Ni(acac)(P(C_6H_5)_3)_4]NO_3$
acetylacetonatotetrakistriphenylphosphine nickel(II) nitrate

(c) $(NH_4)_4[Fe(ox)_3]$
ammonium trisoxalato ferrate(II)

(d) $W(CO)_3(NO)_2$
dinitrotricarbonyl tungsten(II)

## 13.
(a) $[Pt\{P(C_6H_5)_3\}_4](CH_3COO)_4$
tetrakistriphenylphosphine platinum(IV) acetate

(b) $Ca_3[Ag(S_2O_3)_2]_2$
calcium bisthiosulfato argentate(I)

(c) $Ru(As(C_6H_5)_3)_3Br_2$
dibromotristriphenylarsine ruthenium(II)

## 14.
(a) $[Fe(en)_3][IrCl_6]$
trisethylenediammine iron(III) hexachloroiridate(III)

(b) $[Ag(NH_3)(CH_3NH_2)]_2[PtCl_2(ONO)_2]$
amminemethylaminesilver(I) dichlorodinitrito platinate(II)

(c) $[VCl_2(en)_2]_4[Fe(CN)_6]$
dichlorobisethylenediammine vanadium(III) hexacyanoferrate(II)

## 15.
(a) Pentaammine(dinitrogen)ruthenium(II) chloride
$[Ru(NH_3)_5N_2]Cl_2$

(b) Aquabis(ethylenediamine)thiocyanatocobalt(III) nitrate
$[Co(H_2O)(en)_2(SCN)](NO_3)_2$

(c) Sodium hexaisocyanochromate(III)
$Na_3[Cr(NC)_6]$

## 16.
(a) Bis(methylamine)silver(I) acetate
$[Ag(CH_3NH_2)_2](CH_3COO)$

(b) Barium dibromodioxalatocobaltate(III)
$Ba_3[Co(ox)_2Br_2]_2$

(c) Carbonyltris(triphenylphosphine)nickel(0)
$[Ni(CO)(P\emptyset_3)_3]$

## 17.
(a) Tetrakis(pyridine)bis(triphenylarsine)cobalt(III) chloride
$[Co(pyr)_4(As\emptyset_3)_2]Cl_3$

(b) Ammonium dicarbonylnitrosylcobaltate(-I).
$NH_4\,[Co(CO)_2(NO)]$

(c) Potassium octacyanomolybdenate(V)
$K_3[Mo(CN)_8]$

(d) Diamminedichloroplatinum(II)
$[Pt(NH_3)_2Cl_3]$

## 18.
(a) $[Co(NH_3)_6][CuCl_5]$
(b) $[Pt(pyr)_4][PtCl_4]$
(c) $[Pd(NH_3)_2(P\emptyset_3)_2]$
(d) $[Au(ox)_2]^-$

**19.** More than one valid structure may be drawn for these compounds; they are isomers. Here is one:

(a)

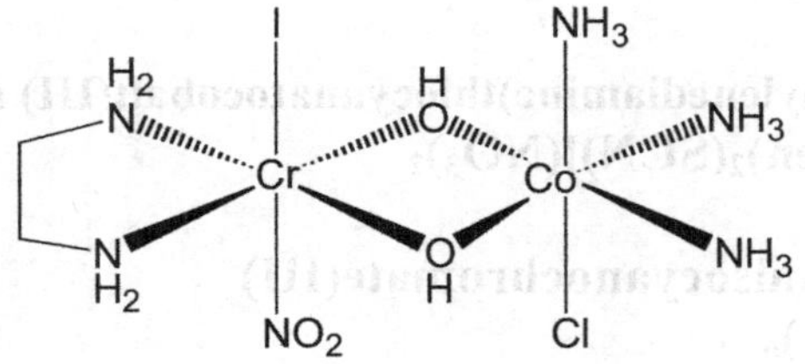

(b)

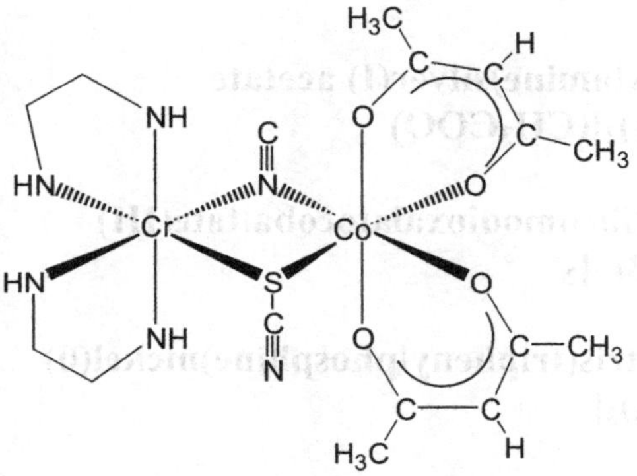

**20.**

(a)

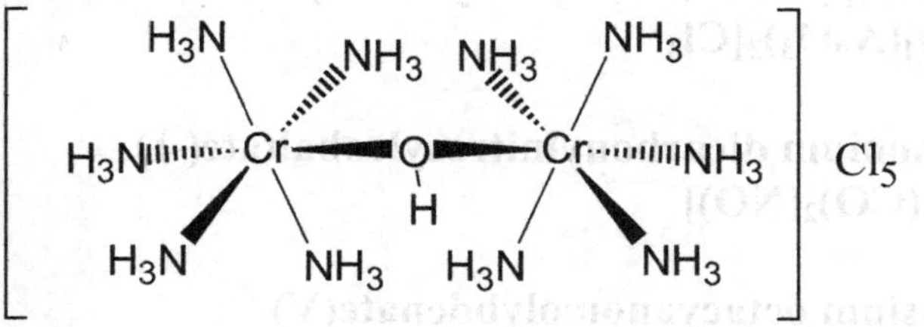

(b)

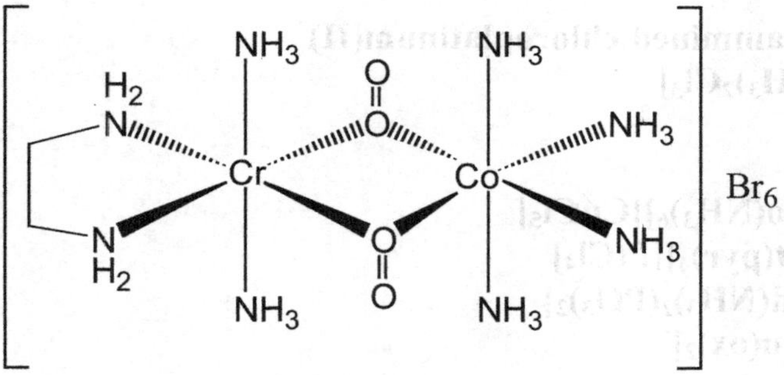

## 21.

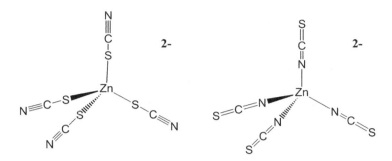

## 22.

(a) dmpm
(b) dppm
(c) dppe
(d) dipamp
(e) triphenyl phosphite
(f) trien

## 23.

(a) $Co_4(CO)_{12}$ = Homobimetallic
(b) $Co_3Rh(CO)_{12}$ = Heterobimetallic
(c) $Mn_2(CO)_{10}$ = Homobimetallic
(d) $Fe_2Ru(CO)_{12}$ = Heterobimetallic

## 24.

(a) $\log \beta_6 = 2.79 + 2.26 + 1.69 + 1.25 + 0.74 + 0.03 = 8.76$
(b) $\beta_6 = 5.75 \times 10^8$

# Chapter 4 Answers to Problems and Exercises

1.

2.

3.

**4.**

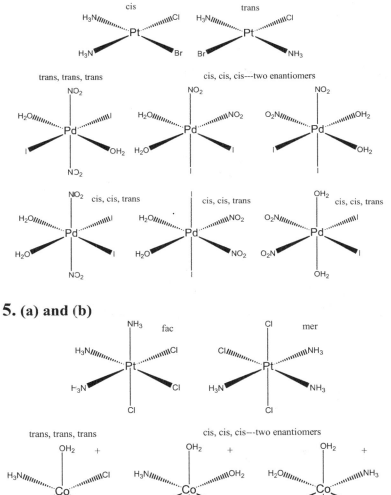

**5. (a) and (b)**

**(c)**

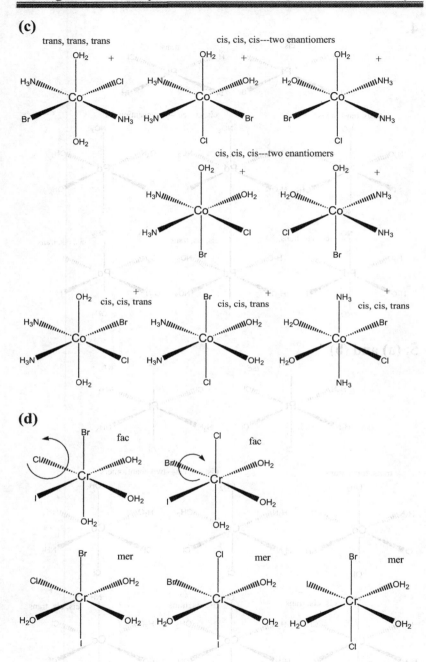

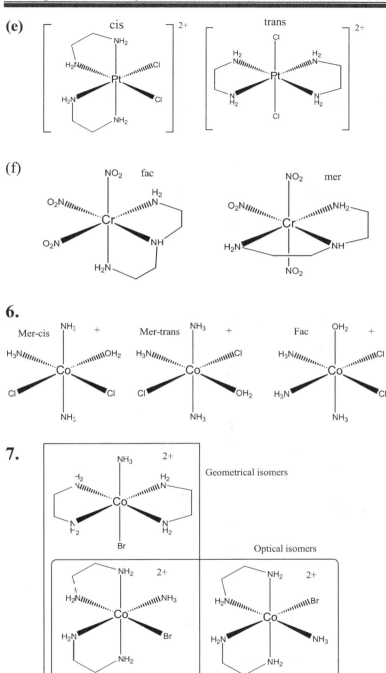

**8.**

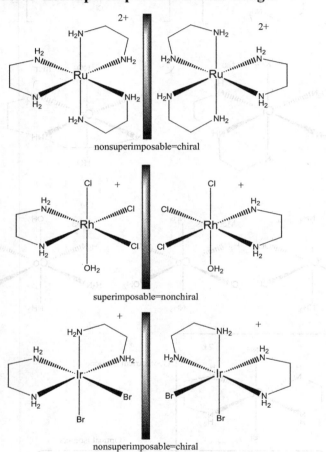

## 9. Chiral=nonsuperimposable mirror images

nonsuperimposable=chiral

superimposable=nonchiral

nonsuperimposable=chiral

**10.**

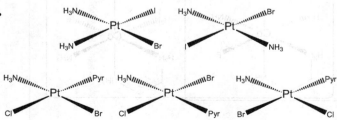

# Inorganic Chemistry: Introduction to Coordination Chemistry

**11.**
Pt 38.8%  or 38.8g x 1mol/195.08g = 0.198 mol Pt
Cl 14.1%   14.1g x 1mol/35.45g = 0.397 mol Cl
C 28.7%    28.7g x 1mol/12.04g = 2.39 mol C
P 12.4%    12.4g x 1mol/195.08g = 0.400 mol P
H 6.02%    6.02g x 1mol/195.08g = 5.95 mol H
  100%  → 100g

thus, dividing by smallest, the empirical formula is:
Pt  0.198/0.198 = 1 Pt atom
Cl  0.397/0.198 = 2 Cl atoms
C   2.390/0.198 = 12 C atoms    $Pt(P(ethyl)_3)_2Cl_2$
P   0.400/0.198 = 2 P atoms
H   5.95/0.198 = 30 H atoms
These must be cis and trans isomers of square planar $Pt^{2+}$

**12.** (c) geometric isomers

**13.** (b) nonsuperimposable mirror images with identical chemical formulae and the same chemical reactivities.

**14.** (c) linkage isomers.

**15.** only one

**16.**

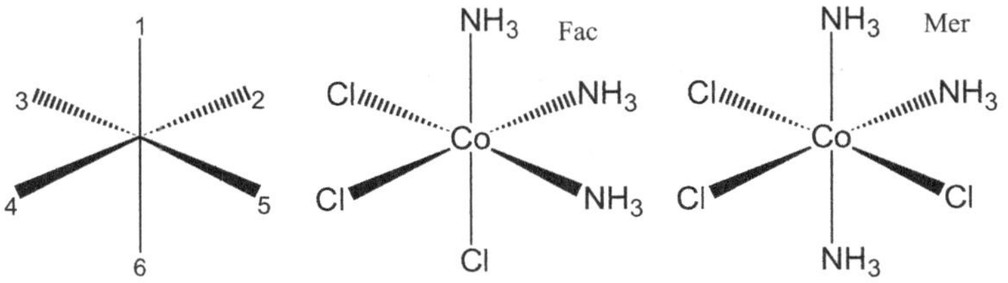

**17.** trans

## 18.

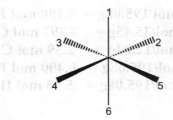

cis, cis, cis—two enantiomers

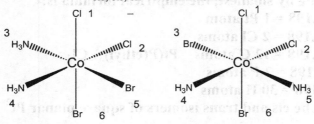

[Co(5,6-dibromo)(1,2-dichloro)(3,4-diammine)]

[Co(3,6-dibromo)(1,2-dichloro)(4,5-diammine)]

## 19.

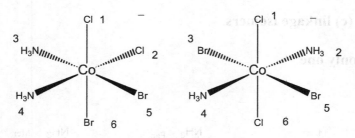

[Co(5,6-dibromo)(1,2-dichloro)(3,4-diammine)]

[Co(3,5-dibromo)(1,6-dichloro)(2,4-diammine)]

# Chapter 5 Answers to Problems and Exercises

**1.** (d) colorless

**2.** (e) Cu+ and $Zn^{2+}$

**3.** (c) both $d^6$

**4.** (a) $d^8$ (b) $d^8$ (c) $d^6$ (d) $d^7$ (e) $d^3$

**5.**  $[Cr(CN)_6]^{4-}$  and  $[Cr(OH_2)_6]^{2+}$
2 unpaired electrons    4 unpaired electrons

Strong Field Case    Weak Field Case

**6.** $[MnCl_4]^{2-}$ has a $Mn^{2+}$ metal center. The Mn atom has a $d^5$ electron configuration. Since Cl⁻ is considered to be a weak field ligand and the tetrahedral geometry is always the weak field case, then the compound would have five unpaired electrons.

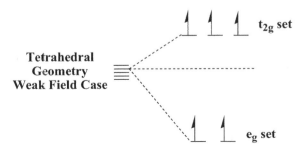

Tetrahedral Geometry Weak Field Case

## 7. Lowest d-orbital splitting energy to the highest.
$[CoCl_6]^{3-} < [Co(ox)_3]^{3-} < [Co(OH_2)_6]^{3+} < [Co(NH_3)_6]^{3+} < [Co(CN)_6]^{3-}$

## 8. Lowest d-orbital splitting energy to the highest.
$[FeCl_4]^{2-} < [FeCl_6]^{4-} < [Fe(OH_2)_6]^{2+} < [Fe(NH_3)_6]^{2+} < [Fe(CN)_6]^{4-}$

## 9. (a) Ni (II)

## 10. (a) the complex is high-spin.

## 11.

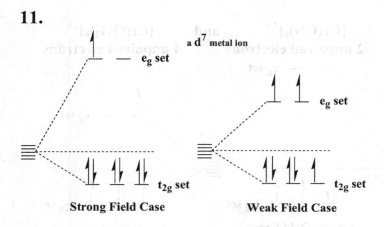

## 12. An octahedral complex is high-spin when the value of $\Delta_o$ is **less** than the energy required to **pair** the electrons.

## 13.

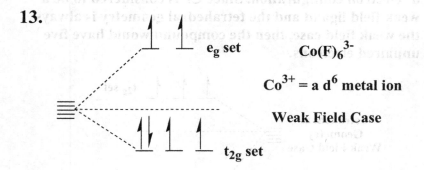

$[CoF_6]^{3-}$   $\mu_{Obs} = 5.3$   This complex is high spin.

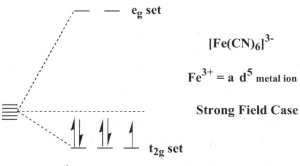

$[Fe(CN)_6]^{3-}$ $\mu_{Obs} = 2.3$   This complex is low spin.

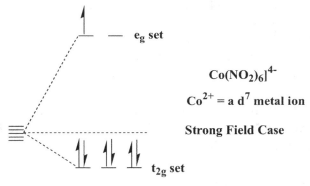

$[Co(NO_2)_6]^{4-}$ $\mu_{Obs} = 1.8$   This complex is low spin.

**14.** $[Cr(H_2O)_6]^{3+}$

**15.**

## 16.

[Co(NH$_3$)$_6$]$^{3+}$ appears orange-yellow  absorbs blue
[Co(NH$_3$)$_5$Cl]$^{2+}$ appears purple  absorbs yellow
[Co(NH$_3$)$_5$(NCS)]$^{2+}$ appears orange  absorbs blue-green
[Co(NH$_3$)$_5$(H$_2$O)]$^{3+}$ appears red  absorbs green

Using the wavelengths of the colors absorbed as an inverse relationship to the energy of $\Delta_o$, then the compounds would have a ranking of from low energy to high energy $\Delta_o$ due to the colors absorbed. These colors are in the order: yellow<green<blue-green<blue. Thus the rankings are:
[Co(NH$_3$)$_5$Cl]$^{2+}$ absorbs yellow <
[Co(NH$_3$)$_5$(H$_2$O)]$^{3+}$ absorbs green<
[Co(NH$_3$)$_5$(NCS)]$^{2+}$ absorbs blue-green<
[Co(NH$_3$)$_6$]$^{3+}$ absorbs blue

Therefore the ligands field strength would be ranked as: Cl- < H$_2$O < NCS < NH$_3$. This matches the order expected in the spectrochemical series.

**17.** [Cr(Br)$_6$]$^{3-}$ < [Cr(H$_2$O)$_6$]$^{3+}$ < [Cr(NH$_3$)$_6$]$^{3+}$

## 18.
The spectrochemical series lists the field strength of the ligands in the order: Cl$^-$ < en < CN$^-$, so the metal complexes, which all have Cr(III), have a decreasing $\Delta_o$ in the order:
[Cr(CN)$_6$]$^{3-}$ > [Cr(en)$_3$]$^{3+}$ > [Cr(Cl)$_6$]$^{3-}$. The energy of the visible light absorbed would follow the same order.

**19.** (b) Cl-

**20.** Jahn-Teller distortion may occur because of the degeneracy in the $e_g$ set.

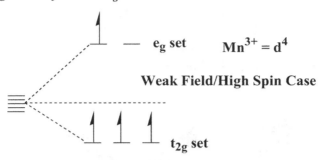

**21.** Jahn-Teller distortion could occur because of the degeneracy in the $e_g$ set.

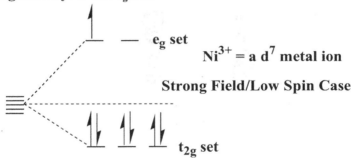

**22.** $Cu^+$ compounds are, by necessity, $d^{10}$ metal centers, which cannot have d-d electron transitions since the d-subshell is filled. On the other hand $Cu^{2+}$ compounds have a $d^9$ metal center, and d-d transitions in the visible range may occur.

**23.** Least amount of CFSE.

**24.** Greatest amount of CFSE.

# KEY to Practice Quiz Chapter 5
**1.**
[CoCl$_6$]$^{3-}$ = **(III)** and it has **six** d electrons
[Cr(CN)$_6$]$^{4-}$ = **(II)** and it has **four** d electrons
[CuCl$_4$]$^{2-}$ = **(II)** and it has **nine** d electrons
[Cr(OH$_2$)$_6$]$^{2+}$ = **(II)** and it has **four** d electrons

**2.**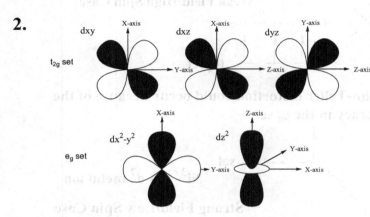

**3.** [W(CN)$_6$]$^{4-}$ 2 unpaired electrons,
[W(OH$_2$)$_6$]$^{2+}$ = 4 unpaired electrons.

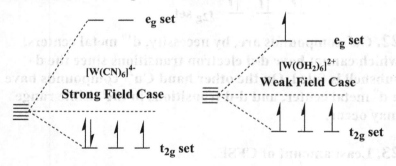

**4.** (A) 1

**5.** (C) [RhCl$_6$]$^{3-}$

**6.** (B) [Os(CN)$_6$]$^{4-}$

**7.** d$^1$ through d$^3$ and then d$^8$ through d$^{10}$

## 8.
(A) the complex is high-spin.
(D) the complex is paramagnetic
(F) $\Delta_o$ is smaller than the pairing energy.

## 9. (E) 4

## 10. FALSE

## 11. $[Tc(CN)_6]^{3-}$ has 2 unpaired electrons, $[TcCl_6]^{2-}$ has 3 unpaired electrons.

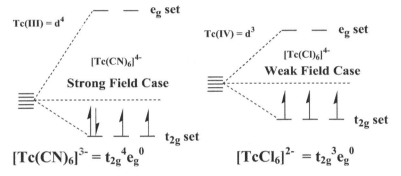

**BONUS question:** Tetrahedral geometry would be paramagnetic, only square planar is diamagnetic.

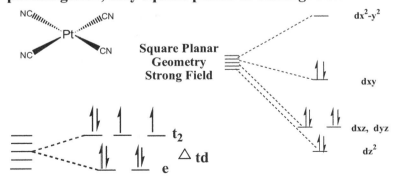

269

# Chapter 6 Answers to Problems and Exercises

1.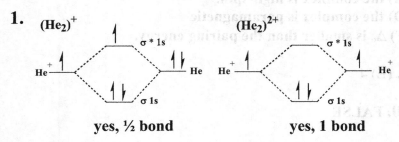

yes, ½ bond     yes, 1 bond

2.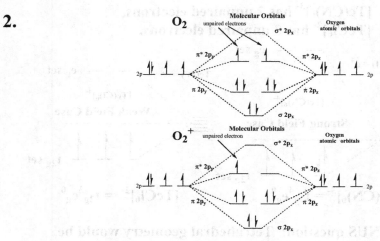

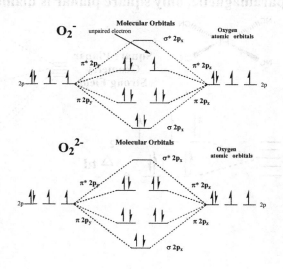

# Inorganic Chemistry: Introduction to Coordination Chemistry

3.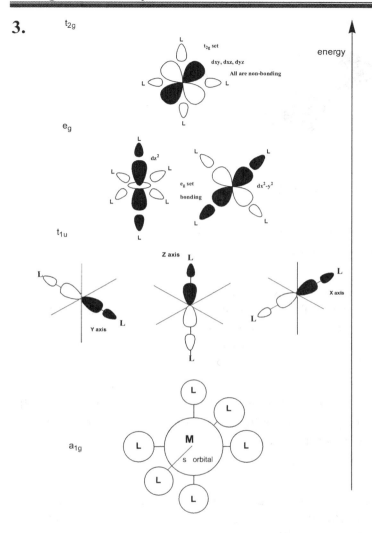

4. The $a_{1g}$, the $t_{1u}$, and the $e_g$ set are all bonding since there are effective overlaps, but the $t_{2g}$ set is nonbonding.

5. Scheme shown on page 159.

6. Scheme shown on page 160.

7. Scheme shown on page 161.

**8.** Dynamic synergism bonding

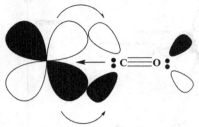

**9.**  M—C≡O:  ⟷  M=C=O:

**10.** Iodide is a larger π-donor ligand
**11.** σ-donor ligand and π-acceptor

**12.** The water molecule has an oxygen atom with more than one pair of electrons on it, so it can function as a π-donor ligand, which decreases the field strength. Also, ammonia is a stronger Lewis base and a better sigma donor than water.

**13.** The o-phenanthroline (phen) ligand has the possibility of forming π-acceptor bonding interactions with the metal, whereas ammonia can not.
**14.** trimethyl phosphite

**15.** The compound must exhibit the structure shown in order for the alkene to use its pi* orbitals to effectively backbond to the filled platinum d-orbitals.

# Chapter 7 Answers to Problems and Exercises

**1.** [V(CO)$_6$], [Cr(CO)$_6$], [Fe(CO)$_5$], [Ni(CO)$_4$].
All of these compounds satisfy the eighteen electron rule, the noble gas formalism, except for the vanadium compound. Vanadium would have a total of five electrons, and with the twelve electrons from the carbon monoxide ligands the total would be therefore 17e-'s, and thus vanadium carbonyl would be a paramagnetic compound.

**2.** With an uneven number of valence electrons the metals form a metal-metal bond in order to satisfy the octet rule and become diamagnetic, and not remain paramagnetic with one unpaired electron.

**3.**

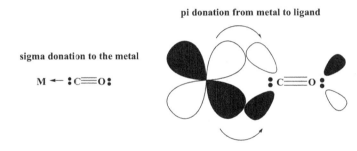

**4.**
Terminal,     bridging,     multiple metal bridging,  semi-bridging

**5.** The more negative charge on the metal center (or the more electron density on the metal center) results in increased backbonding from the metal to the ligand. The more backbonding to the ligand from the metal center means that the antibonding molecular orbitals on the carbon monoxide ligand become more populated, which results in decreased bonding between the carbon and oxygen atoms in the CO molecule. When the bond order is decreased from three to somewhere between three and two, then the infrared stretching frequency is reduced as well. Thus, the $V(CO)_6^-$ molecule has the most negative charge on the metal, the most backbonding, and therefore the lowest stretching frequency. The $Mn(CO)_6^+$ molecule has the most positive charge on the metal, the least backbonding, and therefore the highest stretching frequency.

**6.** (a) $Fe_2(CO)_9$,

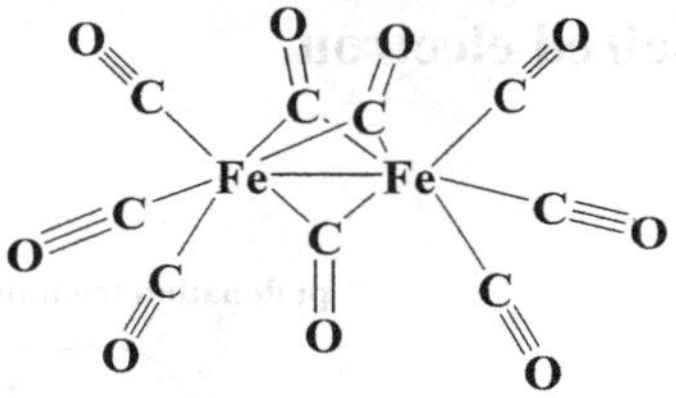

(b) $Ru_3(CO)_{12}$

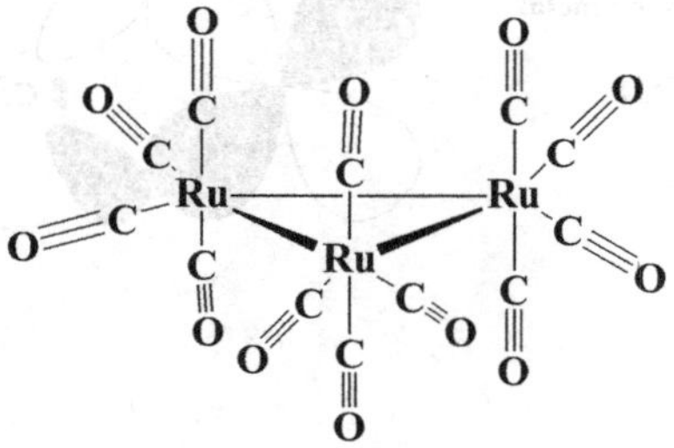

(c) $Rh_4(CO)_{12}$

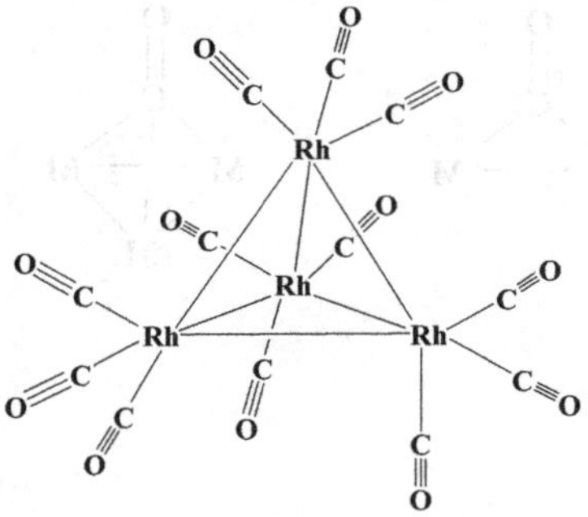

**7.**

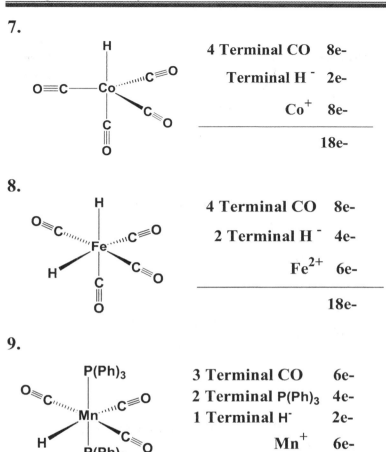

| | |
|---|---|
| 4 Terminal CO | 8e- |
| Terminal H⁻ | 2e- |
| Co⁺ | 8e- |
| | 18e- |

**8.**

| | |
|---|---|
| 4 Terminal CO | 8e- |
| 2 Terminal H⁻ | 4e- |
| $Fe^{2+}$ | 6e- |
| | 18e- |

**9.**

| | |
|---|---|
| 3 Terminal CO | 6e- |
| 2 Terminal P(Ph)$_3$ | 4e- |
| 1 Terminal H⁻ | 2e- |
| Mn⁺ | 6e- |
| | 18e- |

**10.** The Mn group has an odd number of valence electrons, seven, and thus it needs to form a metal-metal bond (or become oxidized to-for example Mn+) in order to achieve eighteen electrons. The compound Mn(CO)$_6$ for example would have a total of 19-electrons. Also, Mn(CO)$_5$ would have a 17-electrons center. By forming a metal-metal bond, and coordinating five carbonyls each, the dimer would achieve 18-electrons around each metal.

## 11.

**(a)**

| | |
|---|---|
| 2 Triphenyl phosphite | 4e- |
| 4 Terminal CO⁻ | 8e- |
| Mo | 6e- |
| | 18e- |

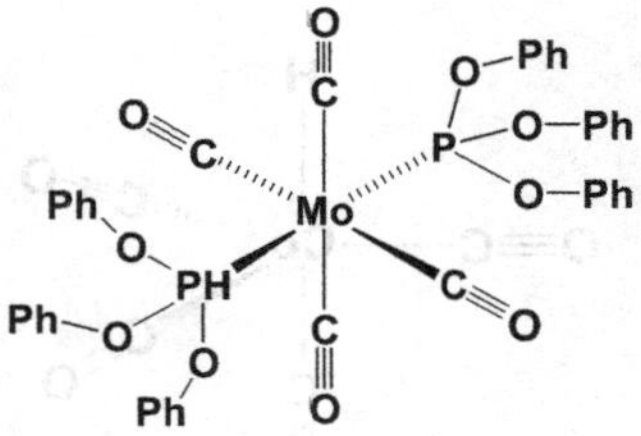

**(b)**

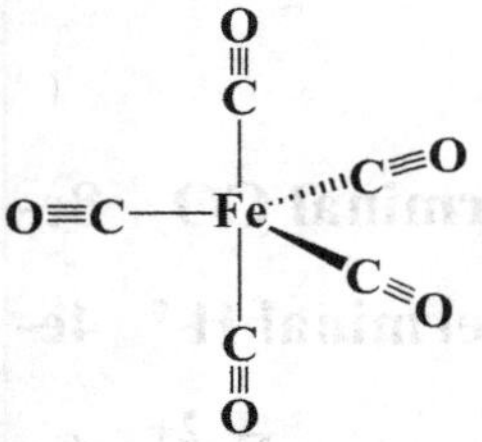

| | |
|---|---|
| Fe | 8e- |
| 5 terminal CO | 10e- |
| | 18e- |

**(c)**

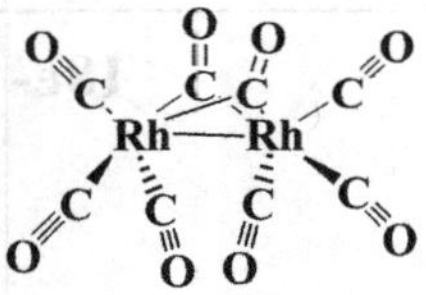

| | |
|---|---|
| Rh | 9e- |
| M-M bond | 1e- |
| 3 terminal CO | 6e- |
| 2 bridging CO | 2e- |
| | 18e- |

**(d)**

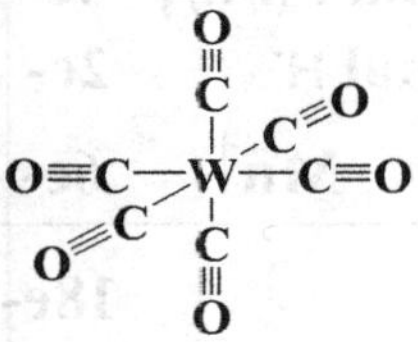

| | |
|---|---|
| W | 6e- |
| 6 terminal CO | 12e- |
| | 18e- |

**(e)**

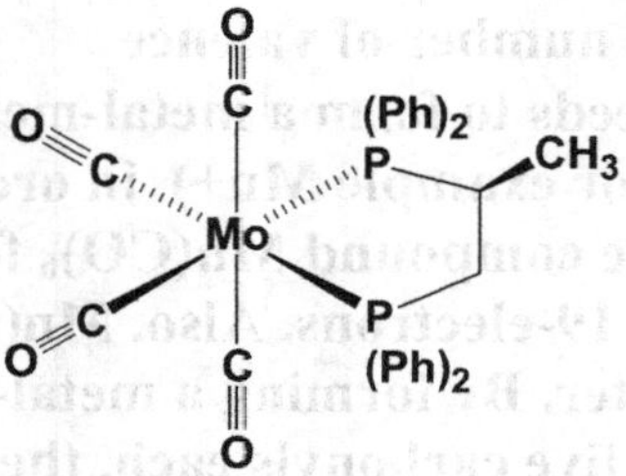

| | |
|---|---|
| PROPHOS | 4e- |
| 4 Terminal CO⁻ | 8e- |
| Mo | 6e- |
| | 18e- |

**(f)**

**Ferrocene**

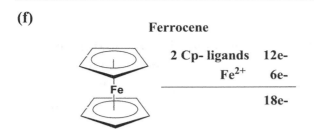

| | |
|---|---|
| 2 Cp- ligands | 12e- |
| $Fe^{2+}$ | 6e- |
| | 18e- |

**(g)**

**Wilkinson's Catalyst**

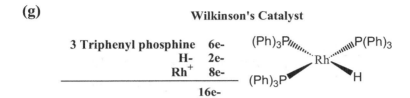

| | |
|---|---|
| 3 Triphenyl phosphine | 6e- |
| H- | 2e- |
| $Rh^+$ | 8e- |
| | 16e- |

**(h)**

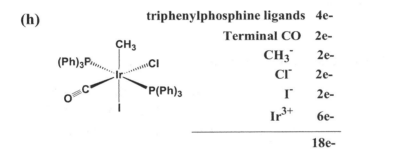

| | |
|---|---|
| triphenylphosphine ligands | 4e- |
| Terminal CO | 2e- |
| $CH_3^-$ | 2e- |
| $Cl^-$ | 2e- |
| $I^-$ | 2e- |
| $Ir^{3+}$ | 6e- |
| | 18e- |

**(i)**

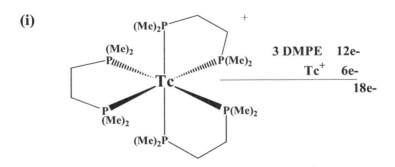

| | |
|---|---|
| 3 DMPE | 12e- |
| $Tc^+$ | 6e- |
| | 18e- |

*two isomers are possible (lambda and delta—see page 108)

**(j)**

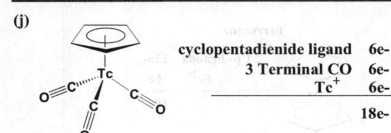

| | |
|---|---|
| cyclopentadienide ligand | 6e- |
| 3 Terminal CO | 6e- |
| $Tc^+$ | 6e- |
| | 18e- |

**(k)**

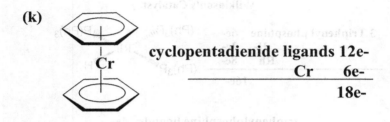

| | |
|---|---|
| cyclopentadienide ligands | 12e- |
| Cr | 6e- |
| | 18e- |

**(l) Each metal has to be counted separately**

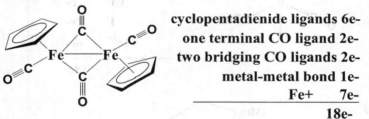

| | |
|---|---|
| cyclopentadienide ligands | 6e- |
| one terminal CO ligand | 2e- |
| two bridging CO ligands | 2e- |
| metal-metal bond | 1e- |
| Fe+ | 7e- |
| | 18e- |

**12.**

Trigonal Bipyramid ⟷ Square Pyramid ⟷ Trigonal Bipyramid

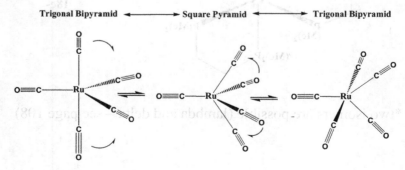

# Inorganic Chemistry: Introduction to Coordination Chemistry

## 13.
(a) Synergism comes from the Greek word *"synergos"* meaning working together. It refers to the interaction between two or more things when the combined effect is greater than if you added the things together.
(b) Increased backbonding reduces CO bond order, decreases the infrared stretching frequency.

## 14.
(a)

$Co_2(CO)_8$

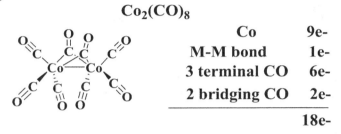

| | |
|---|---|
| Co | 9e- |
| M-M bond | 1e- |
| 3 terminal CO | 6e- |
| 2 bridging CO | 2e- |
| | 18e- |

(b)

Each metal has to be counted separately

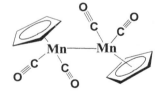

cyclopentadienide ligand 6e-
two terminal CO ligands 4e-
two metal-metal bonds 2e-
Mn+    6e-
18e-

(c)

Each metal has to be counted separately

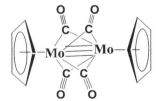

cyclopentadienide ligand 6e-
four bridging CO ligands 4e-
three metal-metal bonds 3e-
Mo+    5e-
18e-

## 15. A quadruple bond is present.

$(Ph)_2P$ ━━━ $P(Ph)_2$
Cl ╲ ╱ Cl
Re ≣ Re
Cl ╱ ╲ Cl
Cl ━━━ Cl

## 16.

| | |
|---|---|
| tropylium$^+$ ligand | 6e- |
| cyclopentadienide ligand | 6e- |
| Cr$^0$ | 6e- |
| | 18e- |

| | |
|---|---|
| benzene ligand | 6e- |
| cyclopentadienide ligand | 6e- |
| Mn$^+$ | 6e- |
| | 18e- |

| | |
|---|---|
| cyclopentadienide ligands | 12e- |
| Fe$^{2+}$ | 6e- |
| | 18e- |

| | |
|---|---|
| cyclobutadiene ligand | 4e- |
| cyclopentadienide ligand | 6e- |
| Co$^+$ | 8e- |
| | 18e- |

| | |
|---|---|
| cyclopentadienide ligands | 12e- |
| Ni$^{2+}$ | 8e- |
| | 20e- |

**17.**

2 Terminal P(Ph)$_3$   4e-
1 Terminal NO+   2e-
1 Terminal Cl⁻   2e-
Ru   8e-
―――――――
16e-

**18.**

3 Terminal CO   6e-
1 Terminal NO+   2e-
Co⁻   10e-
―――――――
18e-

**19.**

2 Terminal CO   4e-
2 Terminal NO+   4e-
Fe$^{2-}$   10e-
―――――――
18e-

**20.**

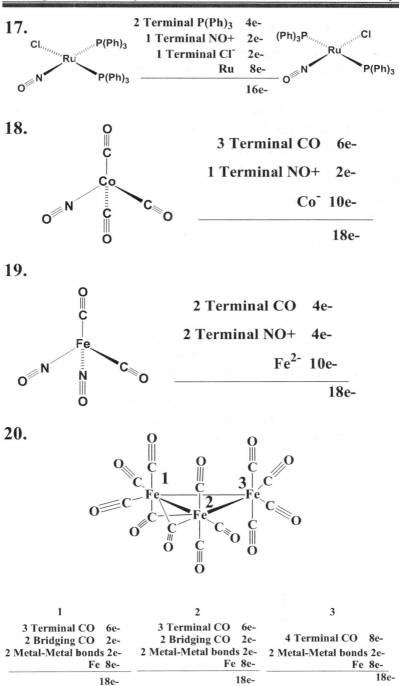

| 1 | 2 | 3 |
|---|---|---|
| 3 Terminal CO   6e- | 3 Terminal CO   6e- | 4 Terminal CO   8e- |
| 2 Bridging CO   2e- | 2 Bridging CO   2e- | 2 Metal-Metal bonds 2e- |
| 2 Metal-Metal bonds 2e- | 2 Metal-Metal bonds 2e- | Fe   8e- |
| Fe   8e- | Fe   8e- | 18e- |
| 18e- | 18e- | |

# Chapter 8 Answers to Problems and Exercises

**1.** Van der Waals forces, or London Dispersion forces are all that hold molecules of carbon dioxide together in solid "dry ice".

**2.** Glass, which is a form of silicon dioxide, can best be described as a <u>Network Covalent</u> type of solid.

**3.** Metallic solids are held together by metallic bonds, which are delocalized electrons that travel through the solid in all three dimension. This non-localized bonding allows for the physical properties of malleability and ductility. Ionic solids are ions---cations and anions---that are held together in a lattice by electrostatic attractions. If the crystallized solid is hit sharply the ions no longer line up and the electrostatic attractions turn into repulsive forces, since alike charged ions would be forced together from the dislocating force of the blow, and the solid cracks apart.

**4.** Which type of metallic lattice is more efficient at packing?
(a) body-centered cubic or <u>face-centered cubic</u>
(b) <u>face-centered cubic</u> or simple cubic
(c) <u>cubic close-packed</u> or body-centered cubic

**5.** Draw a simple band scheme to distinguish a metallic conductor like a Group I metal such as sodium compared to a Group II metal such as calcium.

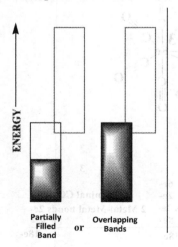

## 6. Draw a simple band scheme to distinguish a metallic conductor like a Group I metal such as sodium compared to an insulator such as diamond.

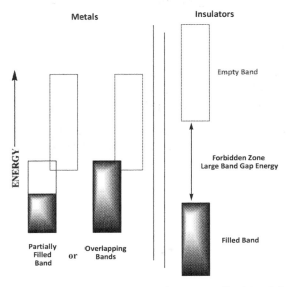

## 7. Draw a simple band scheme to distinguish between an N-type or P-type semiconductor.

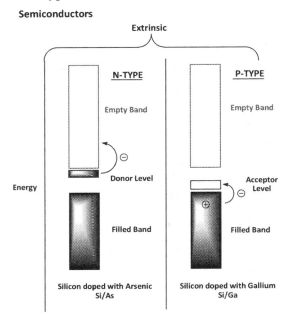

8. (a) arsenic doped germanium = **P-Type**
   (b) boron doped germanium = **N-Type**
9. (a) **gallium** or cesium  (b) **sulfur** or silicon
   (c) **phosphorous** or aluminum  (d) zinc or **germanium**
10. **Increases     Decreases**

11. Carbon is a network covalent solid, and the atoms of carbon are bound together with localized covalent bonds. Silicon is a semimetal, and the bonds are more metallic and delocalized.

12. Classify each of the following compounds as covalent molecular, covalent network, metallic or ionic, based only on their chemical formulas.
(a) $NO_2$ = **covalent molecular**
(b) $CaF_2$ = **ionic**
(c) $SF_4$ = **molecular covalent**
(d) Ag = **metallic**

13. (a) covalent molecular (b) covalent network
    (c) metallic (d) ionic

| Compound type | | Melting Point °C | Boiling Point °C | Electrical Conductor | |
|---|---|---|---|---|---|
| | | | | Solid | Liquid |
| d | RbI | 642 | 1300 | no | yes |
| a | $Se_8$ | 217 | 684 | poor | poor |
| a | $MoF_6$ | 17 | 35 | no | no |
| c | Pt | 1769 | 3827 | yes | yes |
| b | BN | 3000 | --------- | no | no |
| b | B | 2300 | 2550 | no | no |
| d | $CeCl_3$ | 848 | 1727 | no | Yes |
| a | $NO_3F$ | -175 | -45.9 | no | no |
| c | Ti | 1675 | 3260 | yes | yes |
| a | $TiCl_3$ | -25 | 136 | no | no |

14. (a) $PF_3$ = covalent molecular  (b) $PF_5$ = covalent molecular
    (c) $SO_2F$ = covalent molecular  (d) Co = metallic
    (e) LiF = ionic  (f) RbI = ionic  (g) $LuCl_3$ = ionic

15. SC = (8x1/8 of an atom = <u>1 atom</u>)
    BCC = (8x1/8 of an atom + 1 atom = <u>2 atoms</u>)
    FCC = (8x1/8 of an atom + 6x1/2 of an atom + 1 atom = <u>4 atoms</u>)

16. (a) what is the mass of the unit cell?
    (b) What is the volume of the unit cell?
    (c) What would be the density of polonium?

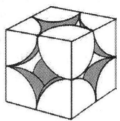

(a) The unit cell only consists of one atom, so the mass of the unit cell would be
(209 g Po /mol Po) x (1 mol Po/$6.022 \times 10^{23}$ atoms) = $3.47 \times 10^{-22}$ g

(b) Since the edge of the unit cell is 336 pm, then the volume would $l \times w \times h$ and we need this in centimeters to get density, so
(336 pm) x (1m/$1 \times 10^{12}$ pm) x (100 cm/m) = $3.36 \times 10^{-8}$ cm

Volume = $l \times w \times h$ = $(3.36 \times 10^{-8}$ cm$)^3$ = $3.79 \times 10^{-23}$ cm$^3$

(c) density = mass/vol = $3.47 \times 10^{-22}$ g/$3.79 \times 10^{-23}$ cm$^3$ = 9.15 g/cm$^3$

17. Calculate the density of sodium metal if it is in a body-centered cubic unit cell with a length of 424 pm.

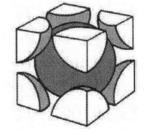

(424 pm) x (1m/$1 \times 10^{-12}$ pm) x (100 cm/m) = $4.24 \times 10^{-8}$ cm

(22.99 g Na /mol Na) x (1 molNa/$6.022 \times 10^{23}$ atoms) = $3.81 \times 10^{-22}$ g

Volume = $l \times w \times h$ = $(4.24 \times 10^{-8}$ cm$)^3$ = $7.62 \times 10^{-23}$ cm$^3$

BCC = 2 atoms, so

density = mass of 2 atoms of Na/vol of unit cell

density = mass/vol = (2 atoms Na)($3.81 \times 10^{-23}$ g Na)/$7.62 \times 10^{-23}$ cm$^3$
= 1.00 g/cm$^3$

**18. The atomic radius of iridium is 136 pm and it crystallizes in a face-centered cubic unit cell. Calculate the density of iridium.**

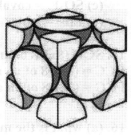

If the radius of the Ir atom is 136 pm, then we can calculate the length of the diagonal since it is the sum of 4 atomic radii.

4 x 136 pm = 544 pm
is the length of the diagonal

h=hypotenuse=diagonal= 4 x radius

Using geometry, since the sum of the squares of the lengths of the edges is equal to the square of the hypotenuse then:

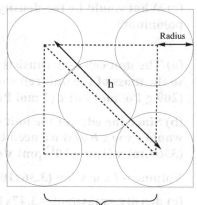

(diagonal length)$^2$ = 2(edge length)$^2$ and then

diagonal length = $(2)^{1/2}$ (edge length) and

edge length=diagonal length/$(2)^{1/2}$ = 385pm   Edge length

(385 pm) x (1m/1x10$^{-12}$ pm) x (100 cm/m) = 3.85x10$^{-8}$ cm

Volume = $l$ x $w$ x $h$ = (3.85x10$^{-8}$ cm)$^3$ = 5.70x10$^{-23}$ cm$^3$

(192.22 g Ir /mol Ir) x ( 1 mol Ir/6.022x10$^{23}$ atoms) = 3.19x10$^{-22}$ g

density =mass unit cell/vol unit cell= 4 atoms Ir/ vol

(4 atoms Ir)(3.19x10$^{-22}$ g/atom Ir) /5.70x10$^{-23}$ cm$^3$ =22.39 g/ cm$^3$

19. The face-centered cubic unit cell would contain 4 atoms of aluminum. If we could calculate the volume of the unit cell we could calculate the mass of the unit cell since we know the density. We can calculate the mass of a single atom of Al, and compare this to the mass of the unit cell which should be 4 times as large.

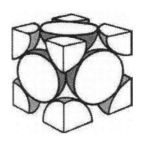

If the radius of the Al atom is 143 pm, then we can calculate the length of the diagonal since it is the sum of 4 atomic radii.

4 x 143 pm = 572 pm is the length of the diagonal

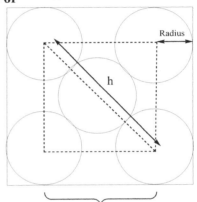

Using geometry, since the sum of the squares of the lengths of the edges is equal to the square of the hypotenuse then:

(diagonal length)$^2$ = 2(edge length)$^2$ and then

diagonal length = $(2)^{1/2}$ (edge length) and

edge length=diagonal length/$(2)^{1/2}$ = 404pm

(404 pm) x (1m/1x10$^{-12}$ pm) x (100 cm/m) = 4.04x10$^{-8}$ cm

Volume = $l$ x $w$ x $h$ = (4.04x10$^{-8}$ cm)$^3$ = 6.59x10$^{-23}$ cm$^3$

(26.98 g Al /mol Al) x (1 mol Al/6.022x10$^{23}$ atoms) = 4.48x10$^{-23}$ g

FCC = 4 atoms, so

density =2.69 g/ cm$^3$ and since density = mass/volume then

mass =density/vol =(2.69 g/ cm$^3$)(6.59x10$^{-23}$cm$^3$) =

1.77x10$^{-22}$ g/ unit cell

the atoms/unit cell = (1.77x10$^{-22}$ g/ unit cell)(1 atom/4.48x10$^{-23}$ g) = 3.95 atoms per unit cell, which is ~4 atoms per FCC unit cell

**Notes**